INTRODUCTION TO THE THEORY AND APPLICATION OF DATA ENVELOPMENT ANALYSIS

INTRODUCTION TO THE THEORY AND APPLICATION OF DATA ENVELOPMENT ANALYSIS

A Foundation Text with Integrated Software

by
EMMANUEL THANASSOULIS
Aston University, Birmingham, United Kingdom.

Distributors for North, Central and South America:
Kluwer Academic Publishers
101 Philip Drive
Assinippi Park
Norwell, Massachusetts 02061 USA
Telephone (781) 871-6600
Fax (781) 871-6528
E-Mail <kluwer@wkap.com>

Distributors for all other countries:
Kluwer Academic Publishers Group
Distribution Centre
Post Office Box 322
3300 AH Dordrecht, THE NETHERLANDS
Telephone 31 78 6576 000
Fax 31 78 6576 474
E-Mail <orderdept@wkap.nl>

 Electronic Services <http://www.wkap.nl>

Library of Congress Cataloging-in-Publication Data

Thanassoulis, Emmanuel.
   Introduction to the theory and application of data envelopment analysis : a foundation
text with integrated software / by Emmanuel Thanassoulis.
      p. cm.
   Includes bibliographical references and index.
   ISBN 0-7923-7429-0
   1. Data envelopment analysis. 2. Industrial efficiency--Measurement. 3. Industrial
productivity--Measurement. I. Title.

HA31.38 .T48 2001
658.5'036--dc21

                                                          2001034441

Printed on acid-free paper.

Printed in the United States of America

To

Meriam, John and Christina

Contents

List of Tables

List of Figures

Preface

1 DATA ENVELOPMENT ANALYSIS

Data Envelopment Analysis (DEA) was initially developed as a method for assessing the comparative efficiencies of organisational units such as the branches of a bank, schools, hospital departments or restaurants. The key feature which makes the units comparable in each case is that they perform the same function in terms of the kinds of resource they use and the types of output they produce. For example all bank branches to be compared would typically use staff and capital assets to effect income generating activities such as advancing loans, selling financial products and carrying out banking transactions on behalf of their clients. The efficiencies assessed in this context by DEA are intended to reflect the scope for resource conservation at the unit being assessed without detriment to its outputs, or alternatively, the scope for output augmentation without additional resources. The efficiencies assessed are comparative or *relative* because they reflect scope for resource conservation or output augmentation at one unit relative to other comparable benchmark units rather than in some absolute sense. We resort to relative rather than absolute efficiencies because in most practical contexts we lack sufficient information to derive the superior measures of absolute efficiency.

DEA was initiated by Charnes Cooper and Rhodes in 1978 in their seminal paper Charnes et al. (1978). The paper operationalised and extended by means of linear programming production economics concepts of empirical efficiency put forth some twenty years earlier by Farrell (1957). In the time since 1978 both theoretical developments and practical applications of DEA have advanced at an 'explosive' pace. For example the book on DEA by Cooper et al. (2000) contains a CD with in their words a 'comprehensive but not exhaustive' list of publications relating to DEA theory or practice, numbering over 1500. Undoubtedly one of the factors driving this proliferation of literature in the field is the fact that DEA brings together theory and practice in a mutually reinforcing and beneficial dynamic. Practical applications of DEA follow theoretical developments in the field while at the same time the applications highlight aspects of practical importance which research must address.

In using DEA in practice we typically go far beyond the computation of a simple measure of the relative efficiency of a unit. Indeed, in some DEA applications we may not do much at all with the numerical measures of comparative efficiency we derive. Far more to the point in using DEA is the building of an understanding of how the transformation of resources to outcomes works. We wish to know what operating practices, mix of resources, scale sizes, scope of activities and so on the operating units may adopt to improve their performance. Benchmark units that may be used as role models for other units to emulate are identified and information is derived on the marginal rates of substitution between the factors of production enriching the picture we have of the units being assessed. More specialised uses of DEA still lead to identifications of types of operating unit which are inherently more efficient by virtue of the kind of unit they are rather than through the operating practices they adopt. Finally more recent developments in the field have shown how to use DEA to measure productivity change over time both at operating unit level and at industry level.

The applications of DEA outlined so far are framed in a production framework where there is some notion of transforming resources to outcomes. Applications of DEA are in fact far broader than the context sketched here would lead us to believe. Resources need not be the traditional types of labour, materials and capital but environmental or contextual. For example in sales outlets the market within which an outlet operates can be treated as a 'resource' which the sales outlet converts into sales revenue. DEA can be used here to assess the relative efficiencies of sales outlets in converting the local market potential to sales revenue of the outlet. Beyond such broad notions of production DEA can be used in contexts totally divorced from production where a choice needs to be made out of a set of multi-attribute alternatives. A typical example is the choice of a site to locate some expensive facility when there is a large number of candidate sites and a multitude of criteria on which the choice must be made. A notional resource of one unit can be applied to all sites and DEA used to assess the relative worth each site yields in terms of the criteria used to choose between the sites.

It is this rich and expanding tapestry both in terms of context and aspects of performance that can be addressed using DEA which accounts for the rapidly expanding literature in this field.

2   THE INTENDED AUDIENCE OF THIS BOOK

The book is aimed at those who wish to learn about the basic principles of how DEA works and how it can be used in practice. The emphasis is on explaining those principles in particular which empower the user to see the potential and the limitations of DEA as an analytical tool and to offer him or her flexibility in adapting the tool to their own needs. The contents of the book represent a distillation of much of the fundamental theory of DEA which is scattered in research papers over a long period of time into a coherent whole beginning from first principles and going up to some of the main extensions of DEA such as its use in assessing productivity change over time and policy effectiveness. Generic DEA approaches are typically presented with simulated real world examples to highlight the use of the concepts involved in practice.

The book is intended to bring the reader up to speed on DEA in as quick and user-friendly a way as possible. It is specifically aimed at those who have had no prior exposure to DEA and wish to learn the essentials of how it works, what its key uses are and the mechanics of using it, including the use of software. Students on degree or training courses or those embarking on research who wish to gain a foundation on DEA should find the book especially helpful. The same is true of practitioners engaging in comparative efficiency assessments and performance management within their organisation themselves or intending to commission such assessments. Examples are used throughout the book to help the reader consolidate the concepts covered.

A limited version of the proprietary *Warwick DEA Software* managed by the author is included with the book to offer the user the possibility of practising with the examples within the book. Restrictions on the use of the software apply as detailed in the license file accompanying the software. The software is upgraded periodically to incorporate additional features reflecting new research developments in the field. For information about obtaining a more up to date or full version of the software and model answers to the practice questions within the book email the author on dea.et@btinternet.com.

3   THE PRESENTATION STYLE OF THE BOOK

The book makes no assumption of familiarity by the reader of DEA, economics or linear programming. Familiarity with basic Algebraic operations is assumed. Background concepts from Production Economics and Linear Programming are introduced within the book as needed. The essentials of Linear Programming as used within DEA are introduced in a brief Appendix in Chapter 3. Readers not familiar with Linear Programming are strongly advised to read that Appendix as background to the numerous Linear Programming models within the book.

The approach adopted for presenting concepts can be said to be from specific to generic. That is concepts are first introduced using simple examples referring to contexts which motivate the concept being presented. Where the context permits graphs are used to demonstrate the use of the concept involved. Once the simple example and/or graph has introduced the concept the generic use of the concept is presented. At this stage notation is general rather than specific to an example. Any underlying technical concepts or proofs are delegated to appendices or referenced in the original research papers and books for the interested reader to follow up.

Chapter 1 is self-contained and introductory in nature. Chapters 2-5 inclusive cover the basics of DEA and can be seen as an integrated whole. At the end of Chapter 5 the reader is ready to engage in using the 'standard' or 'basic' DEA models. Each one of chapters 5-9 is-self contained and presents one or more extensions to the basic DEA models covered in Chapters 2-5. The reader can therefore read selectively from chapters 6-9 inclusive depending on the extension of DEA they are interested in.

4   ACKNOWLEDGEMENTS

I am grateful to an anonymous referee for helpful comments on an earlier draft of the book. I am also grateful to Stefan Puskas for doing so much to get the book to camera ready form. I am most deeply indebted to Bill Cooper not only for being one of the fathers of DEA and for contributing so much to the field since that time but also for offering very helpful comments on a draft of this book. Any errors or omissions in the book are entirely my responsibility. I would be grateful if people would bring any such errors to my attention on dea.et@btinternet.com.

Finally I would like to offer thanks to Professor Robert Dyson of Warwick University for pointing me many years ago in the DEA direction, to my PhD students and academic colleagues at Aston and Warwick Universities who have contributed to our joint work in DEA from which this book draws and to Keith Halstead and Mike Stelliaros of Warwick University for doing so much to code the *Warwick DEA Software* which has proved such an invaluable tool for so many in DEA. I am very grateful to the University of Warwick, for permission to incorporate a limited version of *Warwick DEA Software* with the book.

Emmanuel Thanassoulis, April 2001.

Abbreviations

ADMU	Anchor DMUs
AR I	Assurance Region Typ I
AR II	Assurance Region Typ II
CAPEX	Capital Expenditure
CMT	Cost Malmquist Type Index
CRS	Constant Returns to Scale
DEA	Data Envelopment Analysis
DM	Decision Maker
DMU	Decision Making Unit
DRS	Decreasing Returns to Scale
FDEF	Full Dimensional Efficient Facets
IRS	Increasing Returns to Scale
LHS	Left-hand-side
LINDO	Linear Programming Software
LP	Linear Programming
Max	Maximise
MI	Malmquist Index
Min	Minimise
MPSS	Most Productive Scale Size
NFDEF	Non-Full Dimensional Efficient Facet
NPV	Net Present Value
OLS-regression	Ordinary Least Squares Regression
OPEX	Operating Expenditure
PPS	Production Possibility Set
RHS	Right-hand-side
UDMU	Unobserved Decision Making Unit
VRS	Variable Returns to Scale
WR	Weight Restriction

Chapter 1

INTRODUCTION TO PERFORMANCE MEASUREMENT

1.1 INTRODUCTION

This book is on *Data Envelopment Analysis* (DEA). In its most traditional form DEA is one of the methods which we can use to assess the *comparative efficiency* of homogeneous operating units such as schools, hospitals, utility companies or sales outlets. In such contexts resources and/or environmental factors are converted into useful outcomes and DEA helps us to measure the comparative efficiency with which individual units carry out this transformation process. In less traditional contexts DEA can be used to choose from a set of competing multi-attribute alternatives such as selecting a most preferred site for locating some major facility or a sales outlet.

For simplicity we restrict the introduction in this chapter to the more traditional contexts where DEA is used, namely those contexts where some sort of transformation is carried out by the units being assessed. The chapter is a preamble to the fuller coverage of the theory and use of DEA, to be found in the remainder of the book. It outlines performance measurement as an instrument of management and gives a largely non-technical overview of the types of performance measurement method available, including DEA. It concludes with an illustrative outline of some major sectors of activity where DEA has been used.

1.2 WHY MEASURE PERFORMANCE?

Clearly a principal objective of performance measurement is to enhance various notions of efficiency. In its popular manifestation performance measurement leads to league tables but within the context of performance management that is only the starting point of an exercise in performance measurement. Further detailed analysis and possibly inspection of the best and of the worst performers is then necessary in order to understand the production process and derive useful information which may help both the worst and the best performers to make further improvements in efficiency.

The type information derived from an assessment of performance depends on the aims of the assessment and on the particular assessment method used. It can include but may not be limited to the following:
- Identification of good operating practices for dissemination;
- Most productive operating scale sizes;
- The scope for efficiency savings in resource use and/or for output augmentation;
- Most suitable role model operating units an inefficient unit may emulate to improve its performance;
- The marginal rates of substitution between the factors of production and;
- Productivity change over time by each operating unit and by the most efficient of the operating units at each point in time.

DEA is one of the methods of performance measurement which supports this type of information.

It can be argued that in the broadest sense the measurement and publication of performance data is intended to secure *control* of an organisation. Figure 1.1 (adapted from Dyson 2001) depicts how performance measurement serves the aim of controlling an organisation. It is assumed that the stakeholders of the organisation hold some broad consensus concerning its aims which are reflected in its mission statement. The mission statement is then operationalised through a set of supporting objectives. A set of performance measures are linked to the objectives and through them to the mission statement. Targets may be set for each performance measure reflecting the priorities of the organisation. Performance measures and targets, if used formatively, can bring about a behavioural response in the organisation and this in turn can lead to a change in performance. Thus the change in performance should be broadly in the direction reflected in the mission statement and the priorities of the organisation.

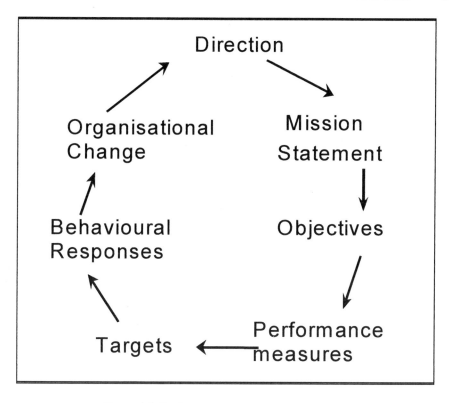

Figure 1.1. Performance Measurement and Control

Performance measurement, and especially the publication of the results in a public sector setting, can unfortunately have unintended consequences too. Smith (1995) for example has identified several types of unintended consequences in this context. They include:

- tunnel-vision through concentration on areas which are the subject of measurement, to the exclusion of other important areas;
- sub-optimisation through the pursuit by managers of their own narrow objectives at the expense of strategic co-ordination;
- Myopia through concentration on short term issues to the exclusion of long term criteria;
- convergence through an emphasis on not being exposed as an outlier on any measure and invite closer scrutiny by the controllers;
- misrepresentation in the sense of manipulation of the data exploiting say definitions of data or loopholes to create a more favourable image than is warranted by the true situation on the ground;
- gaming, such as for example attaining performance below the level possible to forestall the possibility of being set even higher performance targets the following year;

- and ossification through a disinclination to experiment with new and innovative methods because of the risk these involve that in the short term they may lead to poor performance.

Thus with performance measurement as with much else in life, we have costs and benefits to consider. The cost (quite apart from the direct cost of measuring performance) is to be found in the unintended consequences of performance measurement and publication. The benefit is to be found in the control of organisations in the sense of steering them in intended directions and in improved productivity. So long as we desire the benefits of performance measurement we should seek to put in place performance measurement and publication systems which minimise their costs and maximise their benefits.

1.3   PERFORMANCE MEASUREMENT METHODS IN OUTLINE

1.3.1 Unit of assessment

Every instance of comparative performance measurement begins with the implicit, if not explicit, definition of the *unit of assessment*. The unit of assessment is the entity we propose to compare on performance with other entities of its kind. Such entities can for example be schools, bank branches or self-contained functions such as the water distribution function of water companies. The unit of assessment uses a set of resources referred to as *inputs* which it transforms into a set of outcomes referred to as *outputs*. Environmental factors may affect the transformation process and depending on the direction of their impact they can be incorporated in the assessment as inputs or as outputs. The delineation of the unit of assessment and the identification of the corresponding input-output factors are of decisive significance in assessing performance. If we do not delineate the unit of assessment properly or if we omit some important input or output the assessment will be biased.

The measure of performance we typically use reflects our estimate of a unit's potential for resource conservation or for output augmentation. Thus to measure performance we need to estimate the input and output levels at which a unit could have operated if efficient. To illustrate this process let us assume that we want to assess water companies on their water conveyance or 'distribution' function and that the amount of water delivered is sufficient to reflect all the factors which drive operating expenditure. In this context the unit of assessment is the water distribution function. The unit has one input (i.e. operating expenditure) and one output (i.e. amount of water delivered).

Let us compute the ratio of input to output (i.e. operating expenditure to water delivered) for each company. Then if we assume that we have no economies or diseconomies of scale in water distribution the observed *efficient* unit cost will be the lowest ratio of operating expenditure to water delivered we have computed. We can use this efficient unit cost to measure the 'efficiency' on water distribution of each company.

For example if we find that the efficient unit cost is $50,000 per Megalitre of water delivered per day then some company which reports $60,000 per Megalitre of water delivered per day has efficiency $50,000/$60,000 = 0.8333 or 83.33%. The measure of 83.33% reflects the proportion of the unit cost of $60,000 incurred by the company, which would have been sufficient if the company had operated as efficiently as the company with the lowest unit cost. The scope for efficiency savings at the company is $10,000 or 16.67% of $60,000 per Megalitre of water delivered per day.

Unfortunately matters are not so simple in real life. For example we may have economies of scale in the sense that the operating expenditure per Megalitre of water delivered per day should be lower when a company operates at a larger scale. We have not reflected this consideration in the efficiency measure computed above and so smaller companies would be disadvantaged if we use the same benchmark unit cost to assess large and small companies alike. If we do make the situation more realistic still we will realise that not only water delivered but also the number of properties served, the length of main maintained and the number of pipe bursts repaired all jointly drive the level of operating expenditure incurred. So we need a more sophisticated method to assess performance in real life contexts. We outline below some of the methods typically used for measuring performance.

1.3.2 Performance Indicators

A simple and commonly used method for measuring the performance of an operating unit is that of *Performance Indicators*. A performance indicator is typically a ratio of some output to input pertaining to the unit being assessed. One example of a performance indicator was the ratio of operating expenditure per Megalitre of water delivered which we used above to assess efficiencies in water distribution. Unfortunately as we saw in that example, a single performance indicator of this kind is rarely enough to convey the relative efficiencies of real operating units. Not only may we have economies

or diseconomies of scale but we typically also have multiple inputs and/or multiple outputs characterising the operations of the units being compared. For example in water distribution as noted earlier we could define a number of performance indicators such as:
– Operating expenditure per Megalitre of water delivered per day;
– Operating expenditure per Km of main maintained;
– Operating expenditure per 1000 Properties connected to the water distribution main;
– Operating expenditure per burst of main repaired.

In cases where multiple performance indictors pertain we have no unique benchmark of minimum input to output ratio to be used for measuring the performance of each operating unit. In a multi-input multi-output context performance indicators of the ratio variety do not capture how the multiple inputs affect simultaneously the multiple outputs of the transformation process carried out by the unit being assessed. Ratio style performance indicators can only capture in full a transformation process when a single resource and a single output are involved. Once we move to more realistic contexts which involve multiple inputs and/or multiple outputs we need a *modelling* approach to measuring performance. This is outlined in the next section.

1.3.3 Modelling Methods of Comparative Performance Measurement

The modelling approach to measuring comparative performance attempts to arrive at a fuller understanding, i.e. a model, of the production process operated by the units being assessed rather than simply compute indexes of their comparative performance. There are two types of modelling methods of comparative performance measurement. They are the *parametric* and the *non-parametric* methods.

1.3.3.1 Parametric methods for measuring comparative performance

Parametric methods are best illustrated in contexts where either a single input or alternatively a single output pertains. Thus let us assume that we have defined a set of organisational units to be assessed and that they use a single input x to produce the outputs denoted y_r r = 1...s. At this point we could adopt one of two approaches. We could make within the model to be developed no explicit allowance for any inefficiency in production by the units being assessed, or we could make such an allowance.

We begin with the case where we make no explicit allowance in the model to be developed for any inefficiency by the units being assessed, though of course we do not expect all of the units being assessed will in fact be efficient. In this case a model such as that in (1.1) can be hypothesised linking input and outputs,

$$x = f(\beta, y_1, y_2 \ldots y_s) + \eta \qquad (1.1)$$

where y_r $r = 1 \ldots s$ are the known output levels and β is a set of unknown parameters to be estimated. The term η reflects random noise in that for any given set of observed output levels the unit concerned may not exactly use the input level $f(\beta, y_1, y_2 \ldots y_s)$ because there may be factors outside those in the model, including random events, which can make the observed input level x deviate from what we predict with the function $f(\beta, y_1, y_2 \ldots y_s)$. η is assumed to be *normally distributed* with mean value of zero and to be independent of the actual output levels y_r $r = 1 \ldots s$.

Using *ordinary least squares regression* (OLS regression) on the observed input-output correspondences at the units being assessed the parameters β in (1.1) can be estimated. Then expressing a unit's predicted input level $f(\beta, y_1, y_2 \ldots y_s)$ as a fraction of its observed input level x yields a measure of its 'input' efficiency. The larger the ratio the more efficient the unit. (We would normally expect that the computed value of $f(\beta, y_1, y_2 \ldots y_s)$ would exceed the observed input level x for about half of the units being assessed, and it will be equal or below the observed value of x for the rest of the units.) We can assess 'output' efficiency in a case where the units being assessed have multiple inputs and a single output in an analogous manner to that for input efficiency outlined here.

One main class of methods where we make explicit allowance in the model to be developed for any inefficiency by the units being assessed is that of *Stochastic Frontier* methods. These methods address two of the main criticisms of the approach where no explicit allowance is made in the model for any inefficiency by the units being assessed: that they estimate average rather than efficient levels of input for given outputs, and that they attribute all differences between estimated and observed levels of input to inefficiency. The hypothesised version of (1.1) in a Stochastic Frontier approach would be

$$x = f(\beta, y_1, y_2 \ldots y_s) + v + u \qquad (1.2).$$

Notation in (1.2) is as in (1.1) except that now the difference η between the observed input level x and the 'deterministic' level $f(\beta, y_1, y_2...y_s)$ in (1.1) is decomposed into two terms: the random error term v which is normally distributed and the term $u \geq 0$ which reflects inefficiency. A statistical distribution for the inefficiency term u is assumed (typically *half-normal* or *exponential*) and then the parameters β of the model can be estimated using a variant of OLS regression or using Maximum Likelihood methods. The efficiency of a unit is then estimated as the expected value of u given the observed value of v + u at that unit. (See Coelli et al. 1988 Chapter 8 for an introduction to stochastic frontier methods, and Kumbhakar and Lovell 2000 for a fuller coverage such methods.)

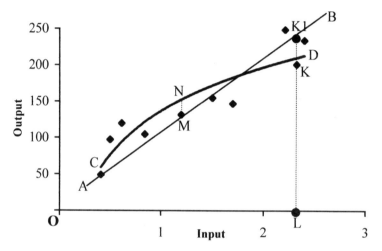

Figure 1.2. An Illustration of Parametric Methods for Efficiency Assessments

Figure 1.2 illustrates graphically the foregoing parametric approaches in the case where the units being assessed use a single input to secure a single output.

OLS regression applied to the input-output correspondences plotted will yield a model which predicts the average level of output that can be secured with a given level of input. If we assume input and output are linked in a linear fashion then OLS regression would predict output levels which lie on some line such as AB. We can adopt the output levels on AB as benchmarks to measure performance. For example we would expect that given its input level unit K should deliver on average the output level at point K1. So its measure of *output* efficiency is LK/LK1 which is the fraction unit K secures of the output level it could have secured on average. Note that apart from the notion of *output* efficiency illustrated in Figure 1.2 we also have the notion of *input* efficiency which measures the potential for input conservation given the output level. The input efficiency measure is illustrated in Figure 1.3. Chapter 2 covers the two measures more fully.

Stochastic Frontier methods applied to the data in Figure 1.2 would estimate a 'maximum' level of output we should expect for each level of input. For some of the units being assessed the maximum output levels estimated could lie *below* their actual output levels because the Stochastic Frontier method assumes there is a random error within the output observed. This error could in principle raise the observed output level by an amount in excess of that by which inefficiency lowers it. Maximum output levels based on the Stochastic Frontier approach could lie on a line such as CD in Figure 1.2. The estimated output efficiency of a unit such as M reflects the difference MN between its observed output and the estimated maximum output feasible for its input level. However, the efficiency measure is not a straightforward ratio of the two as in the OLS regression approach above because the random error (the term v in (1.2)) must be 'netted out' first to arrive at the estimate of efficiency. This process requires the computation of a statistical *expected* value using the assumptions about the statistical distributions of v and u in (1.2).

Compared to the use of performance indicators outlined in the previous section a parametric approach can lead to a better understanding of the production process of the units being assessed and it leads to a summary measure of performance rather than a multitude of performance indicators. However, the approach creates problems of its own. Chief among these is the fact that we have to hypothesise the type of model to be estimated (linear, non-linear, logarithmic etc.). This can lead to a misspecified model. Another problem is that we cannot cope readily with multiple inputs and multiple outputs. These problems are overcome by the non-parametric method of comparative performance measurement outlined next.

1.3.3.2 Non-parametric methods for measuring comparative performance

The main method in this category is DEA, the method of focus for this book. We sketch the approach here using graphical means to convey the key idea of how it works.

We can use the same units as in Figure 1.2 to illustrate how DEA works. The units are plotted in Figure 1.3. In DEA we do not hypothesise a functional form linking input to outputs. Instead, we attempt to construct a *production possibility set* from the observed input-output correspondences at the units being assessed.

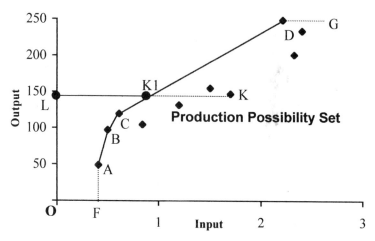

Figure 1.3. An Illustration of an Assessment by Data Envelopment Analysis

The *production possibility set* to be constructed should contain all input-output correspondences which are feasible in principle, including those observed at the units being assessed.

In order to construct the foregoing production possibility set we make certain assumptions, elaborated in Chapter 3. For illustrative purposes we will use three assumptions to construct a production possibility set out of the units plotted in Figure 1.3. The assumptions are:

- *Interpolation* between feasible input-output correspondences leads to new input-output correspondences which are feasible in principle;
- *Inefficient production* is possible;
- The production possibility set is *the smallest set* meeting the foregoing assumptions and containing all input-output correspondences observed at the units being assessed.

Using the interpolation assumption we deduce that all input-output correspondences lying between any two of the units being assessed can in principle be observed. Thus the input-output correspondences along the linear segments AB, BC etc. are feasible in principle. Further, since inefficient production is possible, the horizontal extension from D, DG and the vertical drop from A, AF also contain input-output correspondences which are feasible in principle. By the same token of inefficient production all input-output correspondences to the right and below the piece-wise linear boundary FABCDG are also feasible in principle. Thus we have identified the space consisting of the piece-wise linear boundary FABCDG and the space to the right and below this boundary as the smallest production possibility set satisfying the above assumptions and containing the units being assessed. This will constitute our production possibility set for assessing the units in Figure 1.3. Incidentally we can see now why we call the approach illustrated in Figure 1.3 'data envelopment analysis'. It is because the approach *envelops* the observed input-output correspondences in the course of carrying out an assessment of performance.

Once we have derived the production possibility set in Figure 1.3, to estimate the efficiency of a unit such as K we proceed exactly as we did in Figure 1.2 but using this time the frontier as a reference. For example in the input orientation (i.e. controlling for output level) we draw the horizontal line from K and find that the unit's output level corresponds to a minimum input level at K1. Hence its efficiency is LK1/LK, the fraction to which unit K could in principle lower its input level.

Although for ease of graphical representation we have used a single-input and a single-output example to illustrate DEA, the method works in the same way when multiple inputs and multiple outputs are involved in an assessment. The only difference is that *linear programming* methods (outlined in an Appendix to Chapter 3 and implemented usually within purpose-written computer software) need to be used to construct the production possibility set involved and to the measure the distance of any unit within that space from the efficient boundary.

It is noteworthy that in DEA we arrive at an implicit piece-wise linear frontier (such as ABCD in Figure 1.3) where each linear segment can be thought of as a local approximation to the unknown *efficient* frontier of the production process operated by the units being assessed. (The segments FA and DG are not efficient as we saw earlier.) The data implicitly determines how many different linear segments are needed to approximate the efficient boundary. Thus, unlike parametric methods, the user is relieved of the need to specify a priori the general shape of the boundary (e.g. linear, non-linear, log-linear etc.) and so run the risk of misspecifying it.

The assessment process by DEA yields much more information than the mere assessment of the efficiency of an operating unit. For example we identified that the target input for unit K in Figure 1.3 is at K1 which in turn is based on the interpolation of the performance of units C and D. Identifying units such as C and D with reference to unit K is important. They are on the boundary of the production possibility set and operate relatively efficiently in the sense that we cannot identify other units or combinations of units which dominate them in the sense of yielding more output for given input or use less input for given output. Units C and D can be used as role models an inefficient unit like K can emulate to improve its performance. For benchmark units such as C and D we can also determine whether they operate under economies or diseconomies of scale and get a measure of the economies scale changes can bring about. Further, we can estimate the marginal rates of substitution between the factors of production and these would normally differ along the different linear segments constituting the efficient boundary. So as with the parametric approach earlier, here too we end up with much more information than a mere measure of efficiency.

In fact the DEA methodology has opened up the possibility of addressing performance much more broadly than the initial notion of comparative efficiency measurement might suggest. DEA can be used to address a host of issues which have effect on and are impacted by the productivity of a unit. For example:
- It can be used to decompose efficiency into components attributable to different layers of management or agents involved in the operations of the units being assessed;
- It can be used to assess the impact of policy initiatives on productivity;
- It can be used to measure the change over time in the productivity of the industry as distinct from that of the units operating within it.

Looking at the modelling parametric and non-parametric methodologies outlined here we can say that they take performance measurement onto a new plane. When carried out successfully they model the production process in its totality, leading to an understanding of the transformation processes involved and revealing much more information about the performance of a unit, such as its potential for further attainment, the optimal scale size at which it could operate, the role models it could emulate to improve its performance, the factors or agents responsible for components of under-performance and so on. This is much more information than was ever possible by a mere pairwise inspection of a multitude of often mutually contradicting ratio values in the form of Performance Indicators as still extensively published in the public sector, at least in the UK.

1.4   SOME AREAS WHERE USES OF DEA HAVE BEEN REPORTED

We give here an outline of three areas where comparative performance measurement by means of modelling methods in general, and DEA in particular, has played a role.

1.4.1 Financial Services

Assessment of comparative performance is used extensively in managing the performance of units providing financial services, most notably banks. Berger and Humphrey (1997) for example survey some 130 studies that use 'frontier analysis', including DEA, in banks and other financial institutions. See also Thanassoulis (1999) for a survey of the use of DEA in banking.

Most reported DEA assessments in banking concern *production efficiency* at the branch level. To assess this type of efficiency the bank unit is seen as using inputs such as labour, capital and space to secure outputs such as taking deposits, processing loan and insurance applications and so forth. A smaller number of the reported assessments concern so-called *intermediation efficiency*. From the intermediation perspective, the bank unit is an intermediary collecting funds in the form of deposits and "intermediating" them to loans and other income-earning activities. Intermediation efficiency reflects the branch's effectiveness in converting its labour, capital, space, market potential and so on into sales of its products.

Virtually all analysts begin by determining the efficiency of the unit being assessed and the input and output levels (targets) that would render it relatively efficient, if it is not already so. (The targets for each unit are a by-product of its assessment by DEA as illustrated earlier.)   For example Sherman and Ladino (1995) in an assessment used targets to arrive at an estimate of total potential savings of the order of US $9,000,000 across the branches of the bank they assessed.

Another issue addressed in assessments of banks by DEA is the identification of *efficient peers* for each inefficient unit. (E.g. in Figure 1.3

units C and D were the efficient peers of unit K.) The identification of efficient peers is one of the most valuable outcomes of a DEA assessment. The structure of the DEA model solved means that an inefficient unit and its efficient peers would generally handle a similar mix of products, using a similar mix of resources in similar environmental conditions. This usually makes efficient peers suitable role models for inefficient units to emulate to improve their performance. (Thanassoulis (1997) elaborates on this point more generally in the framework of DEA.) For example Sherman and Ladino (1995) report that they conducted ratio analyses of costs and transactions data and combined the results with field visits to identify differences in operating practices between efficient and inefficient branches. Dissemination of the findings can lead to improved efficiency across the whole system of branches.

One further issue analysts address in DEA applications in banking is that of disentangling the component of the unit's efficiency attributable to its management and that attributable to the *policy* under which the unit operates, which is often not controlled by its management. For example bank branches may be categorised into those with and those without ATM facilities. Each category is known as a policy under which the units operate. At issue is whether the units operating under a given policy are more effective by virtue of that policy. Caution is needed in addressing this question. If the units are assessed by policy group, the efficiency ratings are not comparable between groups since they relate to different benchmark units for each group. If the units are assessed in a single group regardless of the policy under which they operate, the efficiency rating of each individual unit will reflect a combination of the performance of its management and of the impact of the policy under which the unit operates.

In Chapter 7 of this book we cover a DEA-based approach to disentangling managerial from policy effectiveness. In brief the approach involves a two-stage assessment process: In the first stage, the analyst assesses the units for their *managerial* efficiency, each one within its own policy group. This makes it possible to estimate the input and output levels (targets) that would render each unit efficient within its own policy group. The analyst then replaces the observed input and output levels of each unit by its targets, combines the units into a single group, and assesses them afresh. Any inefficiencies identified at this second-stage assessment are attributable to the policy within which a unit operates rather than to its management because the analyst eliminated artificially managerial inefficiency by adjusting the data to within policy efficient levels. This type of analysis reveals the types of branches, or units, that are likely to be efficient and perhaps the environments in which they are most likely to be efficient. Such information is valuable at corporate level, where policies on location, facilities, acquisitions, and divestment of branches are decided.

In sum, modelling methods of performance measurement in the area of financial services provision address a variety of issues such as those outlined above and others such as the impact of size on productivity, productivity changes over time and service quality. This makes them a valuable tool in the management of the performance of units operating in the financial services sector.

1.4.2 Regulation

Regulation of utilities such as gas, water and electricity is another area where comparative performance measurement methods have made substantial inroads. Regulation has since the 1980's moved to a totally new level of prominence both in the UK and elsewhere, largely as a consequence of the privatisations of publicly owned assets undertaken on a massive scale from that time on, worldwide. Despite the declared aim that privatisations should reinforce market competition there remain significant barriers to such competition at least so far as many utilities are concerned. The regulator aims to in effect simulate competition. Using estimates of potential efficiency savings at company level the regulator can factor them into the *determinations* he or she announces. The determinations are normally expressed as limitations to prices for certain services each regulated company can charge.

One case where DEA has featured in the regulated sector is that of OFWAT, the regulator of English and Welsh water companies. In the course of its 1994 Periodic Review of water companies OFWAT used OLS regression and DEA to estimate the potential savings at water companies through improved operating efficiency. Their operations were divided into seven distinct relatively self-contained functions such as *abstraction and treatment of water*, *conveyance of water*, *conveyance of sewage*, *managerial activities* and so on. (See Thanassoulis 2000b for details.) Potential efficiency savings were estimated at function level and were then aggregated to estimate potential savings at company level. Assessment by functions makes it possible to use simpler estimating models since a model for estimating potential savings at company level would require many variables to reflect the totality of company activities. This is unwieldy given the small number of companies under assessment.

To illustrate further the use of modelling methods by OFWAT in its 1994 Periodic Review let us take the function *water distribution*. This is the function of conveying water from treatment works to the clients. In order to assess the comparative performance of the regulated companies on this function during the 1994 review of water companies, OFWAT used OLS regression and DEA. In the OLS regression approach the input was Operating Expenditure or *OPEX*. The outputs were *PMNH* which is the proportion of water delivered 'measured' to non-households, *LENGTH* which is the length of main used in delivering water and *WADELA* which is the estimated amount of water delivered. The OLS regression model derived is in (1.3) where e is the base of natural logarithms.

$$OPEX = 17.84 \ WDELA^{0.61} \ LENGTH^{0.37} \ e^{-1.3 \ PMNH} \qquad (1.3).$$

This formula enables us to predict the operating expenditure level we would on average expect a water company to have for its levels of water delivered, length of main used and proportion of water delivered measured to non-households. This estimate when compared to the company's actual operating expenditure on water distribution makes it possible to rank the companies of cost efficiency.

OFWAT also commissioned during its 1994 review of water companies DEA assessments of performance to be used in support of the assessments conducted using OLS regression. (Outlines of some of these assessments can be found in Thanassoulis 2000a, 2000b and forthcoming.) In the case of water distribution, the DEA assessment used operating expenditure as input and *PROPERTIES, LENGTH* and *WDELA* as output. WDELA and LENGTH are as defined above while PROPERTIES is the number of properties connected to the company's water distribution main. PROPERTIES reflect the number of supply connections served by a company while the LENGTH reflects the dispersion of clients. These two variables capture the scale size of the water distribution network and so we would expect them to influence operating expenditure. The amount of water delivered, WDELA, is a measure of the work done by companies in conveying water and so it too should influence operating expenditure.

The DEA efficiencies in water distribution were computed and the corresponding potential for efficiency savings in expenditure on water distribution was estimated for each company. The potential savings across the full set of companies amounted to £144m on base modelled expenditure of £540.5m in 1992/3 prices, that is 26.67%. (£1 = $1.50 approximately.)

The precise use of the results from the OLS regression and DEA assessments of performance is confidential to OFWAT. However, from published accounts, (see (OFWAT (1995), p. 414)) it is found that OFWAT used the DEA in conjunction with the regression-based results to arrive at a ranking of companies on efficiency. Once the ranks on efficiency were obtained, OFWAT took further factors into account such as the quality of customer service provided by each company and its strategic plans before arriving at the final price determinations (annual price controls) it set.

1.4.3 Police Services

A use of DEA in assessing police services is reported in Thanassoulis (1995). The application was in the context of a wider study of 'crime management' undertaken by the Audit Commission in 1992-1993. Crime management includes a raft of police activities, notably strategies to prevent crime and investigation once it has occurred. The bulk of police effort is applied to investigation and the Audit Commission study focused on police procedures in this area. (The Audit Commission is an independent body set up in the UK in 1982 by Act of Parliament (Local Government and Finance Act 1982) to inter alia encourage the economic, effective and efficient provision of services by publicly funded bodies.)

Unlike most European countries, there is no national police force in England and Wales. At the time of the study policing services were delivered by 43 autonomous *Forces*, eight of which covered the major conurbations — London, West Midlands, Manchester etc. The provincial — that is non-London Forces — varied in size from just under 1000 to 7000 officers in the Force. Thanassoulis (1995) carried out the assessment at Police Force level and excluded the London Metropolitan and the City of London Police Forces. These two Forces were not deemed comparable to other Forces because of their special duties such as the provision of diplomatic protection, which are not generally found at Forces outside London. The remaining 41 Forces were assessed using data for the year 1991.

The assessment used as outcome variables ('outputs') the number of violent, burglary and 'other' crimes cleared, controlling for the number of crimes of each one of these types committed and the number of officers employed by the Force ('inputs'). An *output orientation* was used to estimate how far each Force's clear-ups could rise compared to the benchmark Forces DEA identified. The efficiency ratings were not the only information of interest. There was a desire to understand better the basis of the efficiency rating of each Force, to strengthen the intuitive support for the results obtained and to identify ways Forces may improve their performance.

To this end it was decided to review 'efficient' Forces in two complementary ways:
- The intuitive appeal of the alternative sets of marginal rates of substitution between clear ups identified by DEA and;
- The frequency with which each efficient Force was used as a benchmark for other Forces.

The intuitive appeal of each set of marginal rates of substitution DEA identified was judged with reference to the relative 'worth' DEA was placing on say the clear up of a violent crime relative to that on a burglary. This process led to 'dropping' some of the Forces from the set of efficient Forces because the marginal rates of substitution rendering the former efficient valued some type of clear up disproportionately highly or in a counter-intuitive way relative to some other type of clear up. The subset of genuinely efficient Forces was further modified to include only those Forces which were frequently used as benchmarks to show other Forces as relatively inefficient. Thus the final set of Forces deemed to be performing well were those which used intuitively acceptable trade-offs between clear ups and were benchmarks for a large proportion of the Forces which were not efficient.

The assessment also addressed the issue of how inefficient Forces might improve their performance. One step in this direction was the identification of role model efficient Forces they could emulate. In the event, for each inefficient Force only one of the efficient Forces could act as a role model. This is because there was uneasiness about the comparability between some of the inefficient Forces and their initial set of efficient peers. It was for example felt that the more economically and socially deprived an area is the harder it is to solve crimes for reasons such as many suspects having no fixed address and the general public being unco-operative with the police. This meant an inefficient Force covering substantial relatively deprived areas could not really be compared with those of its efficient peers which did not cover similarly deprived areas.

In order to address this concern 'families' of Forces were used. The Audit Commission had divided Forces into 'families' using a number of socio-economic and settlement pattern indicators so that a consensus existed that Forces in each family cover similar policing environments. Where an inefficient Force did have efficient peers from its own family it was contrasted to peers from within rather than outside its family. Most Forces did have efficient peers from within their family. This is not coincidental. DEA selects efficient peers for an inefficient Force which largely match its input-output mix. That mix in the case of Forces reflects in essence the mix of crimes they have to deal with which crimes in a general sense reflect at least some of the factors used for creating the families of Forces. Where an inefficient Force did not have efficient peers from its own family it was contrasted with efficient Forces from within its family even if they were not its DEA peers. All families had at least one efficient Force and no inefficient Force had more than one peer from its own family.

The identification of inefficient Forces and their efficient peers made it possible to discuss the DEA identifications of weaker and stronger Forces on performance without recourse to the technical detail of DEA but by merely focusing on ratios such as crimes cleared per officer, proportion of each type of crime cleared and so on. Ratios precede DEA as instruments of performance assessment and many practitioners find them easier to use. The strength of DEA over conventional ratio analysis lies in that it takes into account simultaneously all input and output variables rather than one output to input ratio at a time which could lead to misleading conclusions. (See the assessment in the banking context by Sherman and Ladino (1995) referred to above where ratios of inputs to outputs post the identification of efficient and inefficient units by means of DEA were also used to better understand the performance of units.)

Full details on the assessment of police Forces outlined above can be found in Thanassoulis (1995). Another use of DEA in assessing police services can be found in Carrington et al. (1997).

1.5   CONCLUSION.

This book is dedicated to DEA, a method for comparative efficiency assessments in contexts where multiple homogeneous units such as bank branches, retail outlets, schools or hospitals deliver goods and/or services. DEA yields a summary measure of the comparative efficiency of each operating unit and much other information useful in managing the performance of the operating units. Such information includes, but is not limited to, good operating practices which can be disseminated to all operating units, targets of performance a unit may be set, the most productive scale size at which a unit may operate, role model units an inefficient unit may emulate to improve its performance, and the extent to which a unit has improved in productivity over time. Many other uses of DEA exist and others are being identified as research in the area proceeds.

One of the reasons why DEA is seeing so much use is that it requires minimal assumptions about how the factors of production, in the form of 'inputs', and the outcomes of production, 'outputs', relate to each other. Further, assessment by DEA relates to 'best' or 'efficient' rather than average behaviour. Efficient benchmarks are more appropriate in the context of performance measurement.

The remainder of this book explains how DEA works, its advantages, its limitations and how it can be used in practice. Sample software is included with the book and is integrated with the presentation of DEA concepts to help the reader understand and use in practice DEA.

Chapter 2

DEFINITIONS OF EFFICIENCY AND RELATED MEASURES

2.1. INTRODUCTION

DEA is a method for measuring *comparative* or *relative* efficiency. We speak of *relative* efficiency because its measurement by DEA is with reference to some set of units we are comparing with each other. We cannot in general derive by means of DEA some *absolute* measure of efficiency unless we make additional assumptions that the units being compared include a 'sufficient' number of units which are efficient in some absolute sense. Thus in a practical setting units which we may find efficient by DEA may in fact be capable of improving their performance even further.

This chapter introduces the basic concepts of relative efficiency and the associated measures.

2.2. UNIT OF ASSESSMENT AND INPUT-OUTPUT VARIABLES

As we saw in Chapter 1, the definition of the *unit of assessment,* that is the entity whose efficiency we wish to measure relative to that of other entities of its kind, is one of the first steps in carrying out assessments of comparative performance. We shall adopt in respect of the unit of assessment the term *Decision Making Unit* (DMU), coined by Charnes et al. (1978), in their seminal paper on DEA. DMUs should be homogeneous entities in the sense that they use the same resources to procure the same outcomes albeit in varying amounts. For example bank branches are homogeneous as they all perform the same tasks but of course they differ in the relative and absolute levels of activities and resources.

The characterisation of the unit of assessment as "decision making" implies that it has control over the process it employs to convert its resources into outcomes. In DEA the resources are typically referred to as "*inputs*" and the outcomes as *"outputs"* and these are the terms we shall adopt. A DMU transforms inputs into outputs in a process depicted in Figure 2.1.

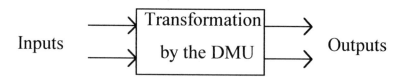

Figure 2.1.   A DMU Transforms Inputs into Outputs

The identification of the inputs and the outputs in an assessment of DMUs is as difficult as it is crucial. The inputs should capture all resources which impact the outputs. The outputs should reflect all useful outcomes on which we wish to assess the DMUs. Further, any environmental factors which impact the transformation of resources into outcomes should also be reflected in the inputs or the outputs depending on the direction of that impact.

In general terms, the essential idea is that we wish to assess how efficiently each DMU is handling the transformation process when compared to other DMUs engaged in that same process. To do this we must capture in the outputs what the DMU achieves, take into account the resources it uses and allow within the input-output set for any factors beyond the DMU's control which impact its performance. What is under DMU control will in general depend not only on the nature of the activities in which a DMU is engaged but also on the decision making authority it has within the broader framework of its parent organisation, if any.

This brief discussion indicates how complex is the issue of input-output variables in DEA assessments and so we shall return to it again in Chapter 5, after we introduce other concepts involved in the measurement of efficiency.

2.3. PARETO EFFICIENCY AND MEASURES OF INPUT AND OUTPUT EFFICIENCY.

Measures of efficiency are based on estimates of the degree to which the DMU concerned could have secured more output for its input levels, or the degree to which it could have used less input for its output levels. Thus at the outset we need to ask whether the DMUs in question have more discretion over input or over output levels. The answer depends on context. For example hospitals have relatively little control over the output levels which would be patients of various categories needing treatment. They would have more control over the input levels, likely to be resources such as doctors, nurses etc. On the other hand schools have little control over the input levels, likely to be measures of the innate ability of the students taught and their socio-economic background, and more control over the outputs likely to be measures of the attainments of students on exit from school.

Depending on whether inputs or outputs are controllable different measures of efficiency are appropriate. Before introducing these measures let us first define *Pareto* efficiency. Two definitions are given, the one labelled *output orientation* is appropriate when outputs are controllable and the one labelled *input orientation* is appropriate when inputs are controllable.

Let a set of homogeneous DMUs use one or more inputs to secure one or more outputs. Then:

Output orientation: A DMU is *Pareto-efficient* if it is not possible to raise anyone of its output levels without lowering at least another one of its output levels and/or without increasing at least one of its input levels.

Input orientation: A DMU is *Pareto-efficient* if it is not possible to lower anyone of its input levels without increasing at least another one of its input levels and/or without lowering at least one of its output levels.

These definitions are stated mathematically in DA2.1.1 in Appendix 2.1.

Two measures of efficiency, relating respectively to the output and input orientation above are most commonly used. They are as follows:

Technical output efficiency:	The technical output efficiency of a DMU is the maximum proportion any one of its observed output levels represents of the level that output takes when all outputs are expanded *radially* as far as feasible, without detriment to its input levels.
Technical input efficiency:	Let all inputs of a DMU be contracted *radially* as far as feasible, without detriment to its output levels. The technical input efficiency of the DMU is the maximum proportion any one of its contracted input levels is of the observed level of that input.

These definitions are stated mathematically in DA2.1.2 and DA2.1.3 in Appendix 2.1.

The measure of output efficiency reflects the extent to which the output levels of the DMU concerned can be raised through improved performance and no additional resource, while maintaining its output *mix* (i.e. radial output expansion). The measure of input efficiency reflects the extent to which the input levels of the DMU concerned can be lowered through improved performance and no output reduction, while maintaining its input mix.

The input mix of a DMU is reflected in the ratio its input levels are to each other. Output mix is defined in a similar manner.

Evidently the measures of efficiency rely on estimating *maximum* output levels for given input levels, or alternatively *minimum* input levels for given output levels. It is these estimates that DEA enables us to make and thereby arrive at measures of input or output efficiency. The reason for the prefix 'technical' to the above efficiency measures will become clear later when it is contrasted with efficiency measures which use input prices.

Figure 2.2 illustrates the difference between input and output oriented measures of efficiency. It depicts the case where DMUs produce a single output using a single input. The curve OD is the locus of maximum output levels attainable for given input levels. OD is thus the *efficient boundary* of the *production possibility set* located between the input axis and OD. (See Chapter 1 for the definition of the production possibility set.) Let us now consider DMU A in relation to the definition of Pareto-efficiency and the measures of efficiency introduced earlier.

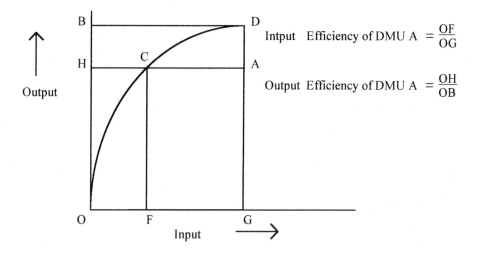

Intput   Efficiency of DMU A $= \dfrac{OF}{OG}$

Output   Efficiency of DMU A $= \dfrac{OH}{OB}$

Figure 2.2.   Measures of Input and Output Efficiency

Clearly DMU A could have been operating at point D to attain maximum output for its input level. Alternatively, DMU A could have been operating at point C to use minimum input for its output level. Thus DMU A is not Pareto-efficient in that it could deliver more output for no additional input or use less input without detriment to its output level. The measure of the technical output efficiency of DMU A as defined earlier is the proportion its observed output is of the maximum level its output could be for its input, i.e. $\dfrac{OH}{OB}$. Similarly the technical input efficiency of DMU A as defined earlier is $\dfrac{OF}{OG}$. OF is the minimum input level at which DMU A could deliver its output, while OG is its observed input level. (We will see later in this chapter a two-input case where the notion of *radial* contraction of inputs comes into play in arriving at a measure of efficiency.)

Note that the orientation in which we measure technical efficiency can impact the result obtained. For example it is evident that in Figure 2.2 DMU A has a different efficiency rating depending on whether we measure its technical input or output efficiency. This will generally be so except when the DMU operates what is known as a *constant returns to scale* technology in which case input and output efficiency measures are equal. (Issues of returns to scale are covered in detail in Chapter 3, suffice it here to say that under constant returns to scale the productivity of a unit cannot be affected by scale of operation.) Finally note that if a DMU is Pareto-efficient it will lie on the efficient frontier and so its input level cannot be lowered without lowering its output level, nor can its output level be raised without raising its input level. Thus its technical efficiency will be 1 whether we measure it in the input or in the output orientation.

2.4. INPUT OVERALL, ALLOCATIVE AND TECHNICAL EFFICIENCIES

So far we have defined efficiency and measures of it with reference to input and output levels without regard to their values. Efficiency defined in this manner is known as '*technical*' efficiency. Because technical efficiency does not reflect relative input prices or output values it can give a false account of the performance of a DMU. For example a DMU which is Pareto-efficient in the technical sense will be using the minimum input levels necessary to secure its output levels <u>given</u> its input mix. The input mix of the DMU may, however, be too costly, depending on the relative prices of the inputs. Thus the DMU may be operating in a very costly, albeit technically efficient way.

Figure 2.3 illustrates how some Pareto-efficient DMUs may be performing better than others in cost terms and how we can measure efficiency so as to reflect performance in terms of costs. It depicts a case where two inputs are used to secure a single output and where efficient production is characterised by constant returns to scale. So output has been standardised to one unit and X_1 and X_2 stand respectively for units of input 1 and input 2 used per unit of output. The space to the right and above the piecewise linear curve BDEF and its vertical and horizontal extensions contains all feasible input levels capable of delivering a unit of output. The piece-wise linear curve BDEF, excluding the vertical and horizontal extensions from B and F respectively, is the locus of Pareto-efficient input levels in that on that curve lowering one input level would require the raising of the other.

Let us now assume that the prices of inputs 1 and 2 are respectively $p_1 = $ \$4 / unit, $p_2 = $ \$10 / unit. Consider now the *isocost* line containing the input level combinations with aggregate costs of \$20. The equation of this isocost line is $4x_1 + 10x_2 = 20$ and its graphical representation is labelled C* in Figure 2.3. The line is tangential to BDEF at E. If the aggregate cost level were to change from \$20 to C' > \$20 the new isocost line would lie to right of and above E while if C' < \$20 the isocost line would not intersect the boundary

BDEF or the space to the right and above that boundary. Thus the point E has the combination of input levels which can secure a unit of output <u>at the lowest aggregate cost feasible</u>. Pareto-efficient point E therefore has superior performance not only to that of inefficient input combinations for delivering a unit of output, but also superior to all other Pareto-efficient points.

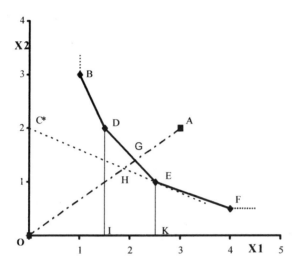

Figure 2.3. Contrasting Technical and Input Allocative Efficiencies

When input prices are known we can measure the *input allocative efficiency* of a DMU, introduced by Farrell (1957) as *price efficiency*. Input allocative efficiency reflects the 'distance' of the input mix used by a DMU from the optimal mix it could have used to minimise the cost of output, in light of input prices. Input allocative efficiency complements the measure of technical efficiency we introduced initially.

The measure of input allocative efficiency can be illustrated with reference to DMU A in Figure 2.3. Were DMU A to become technically efficient under its chosen input mix it would operate at G. Thus (see Appendix 2.2),

$\dfrac{OG}{OA}$ is the *technical input efficiency* of DMU A.

The fraction to which the aggregate cost of the inputs at G could be lowered (see Appendix 2.2) is $\dfrac{OH}{OG}$ and we define,

$\dfrac{OH}{OG}$ the *input allocative efficiency* of DMU A.

The fraction to which the aggregate cost of the inputs at A could be lowered is $\dfrac{OH}{OA}$ and we define

$\dfrac{OH}{OA}$ the *input overall efficiency* of DMU A.

It can be readily deduced from Figure 2.3 that $\dfrac{OH}{OA} = \dfrac{OG}{OA}\dfrac{OH}{OG}$. Thus we have:

Input Overall Efficiency = Technical Input Efficiency × Input Allocative Efficiency.

Generalising from the illustration in Figure 2.3 we have the following definitions of input allocative and overall efficiencies:

Input allocative efficiency:	For given input prices let C_{min} be the minimum cost at which a DMU could secure its outputs and C_{te} the cost of its technically efficient input levels for its input mix. The input allocative efficiency of the DMU is C_{min} / C_{te}.
Input overall efficiency:	Let C_{min} be as above and C_{ob} the cost of the DMU's observed input levels. The input overall efficiency of the DMU is C_{min} / C_{ob}.

A mathematical statement of these definitions can be found in DA2.1.4 in Appendix 2.1.

In practice it is naturally more important that a DMU be overall rather than merely technically efficient. A DMU cannot be overall efficient without also being technically efficient but the converse is not true. Unfortunately, it is not always possible to measure input allocative and therefore overall efficiency because we do not always have suitable input prices. Indeed, in some contexts input prices are meaningless and we have to focus on technical efficiency. A case in point is the measurement of school effectiveness. The notion of a 'price' on the academic attainment of a pupil on entry to school or on pupil socio-economic background, both frequently used as inputs in school value-added studies, is difficult to say the least.

2.5. AN ILLUSTRATION

An organisation has 4 DMUs producing a single product using two inputs. The table below shows the number of units of each input used per unit of output by each DMU.

Table 2. 1.   Units of Input per Unit of Output

DMU	Input 1	Input 2
1	2	3
2	4	1
3	2	2
4	1	4

The DMUs operate in a technology in which if X1 and X2 denote respectively the amount of input 1 and 2 used then one unit of output can be secured so long as a minimum of one unit of each input is used and the input levels satisfy the expression $1.2X1 + X2 \geq 3$.

a) Draw the space of feasible input levels for securing a unit of output. Identify the boundary of the space and comment on the boundary's peculiarities, if any.
b) Using the graph compute the technical input efficiencies of the DMUs.
c) Inputs 1 and 2 cost $6 and $4 per unit respectively. Compute the input allocative and the input overall efficiency of DMU 3.

a) The feasible input levels for securing a unit of output are to the right of and above ABCD in Figure 2.4 and the points where DMU1…DMU4 operate are as indicated. BC is the locus of Pareto-efficient input levels for securing a unit of output. A reduction of one input level on BC can only take place by raising the level of the second input. The parts of the boundary labelled CD and AB are not Pareto-efficient, beyond the points B and C. Any point on AB away from B is dominated by B and any point on CD away from C is dominated by C in the sense that for the same output level B and C use less of one of the inputs than the points they dominate.

b) The technical input efficiencies of DMUs 4 and 2 are 1. This is because they lie on the boundary and in both cases the input levels cannot be reduced *radially (i.e.* maintaining the DMU's input mix) while continuing to secure a unit of output. Both DMUs are *not* Pareto efficient. The technical input efficiencies of the remaining DMUs are as follows:

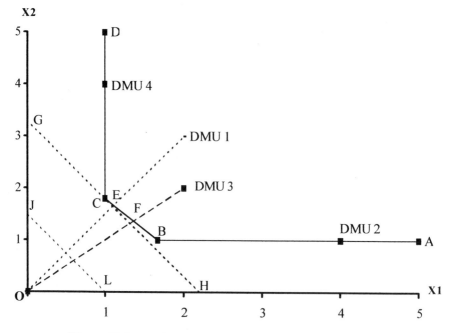

Figure 2.4. Feasible Input Levels for Securing a Unit of Output

DMU 1: The technically efficient levels for its input mix are at E, where the line from the origin to DMU1 intersects the efficient boundary BC. At E we have $X1 = 1.111$ and $X2 = 1.667$. Expressing these as proportions of the observed input levels of DMU 1 we have $\dfrac{1.111}{2} = \dfrac{1.667}{3} = 0.556$. This is the technical input efficiency of DMU 1.

DMU 3: The technically efficient levels for its mix of inputs are at F where $X1 = X2 = 1.364$. Expressing these as proportions of its observed input levels we have $\dfrac{1.364}{2} = 0.682$. This is the technical input efficiency of DMU 3.

c) The isocost line for any given cost level K is $6X1 + 4X2 = K$. For example setting $K = \$6$ we obtain the input combinations on the line JL. None of these combinations can secure a unit of output as JL does not intersect

ABCD or the space above and to the right of this piecewise linear boundary. If we raise K we shall obtain input combinations on lines parallel to JL and closer to ABCD. One such parallel line to JL is GH which is tangential to ABCD at C. Thus out of all feasible input levels on ABCD, above and to the right of it those at C minimise the aggregate cost 6X1 + 4X2 of the inputs. The input levels at C are (1, 1.8) and their aggregate cost is $13.20. The technically efficient input levels for the input mix of DMU 3 are (1.364, 1.364) and their aggregate cost is $13.64. Hence, using the definition given earlier, we conclude that the input allocative efficiency of DMU 3 is 13.20 / 13.64 = 0.967. This is quite close to 1 reflecting the fact that the input mix of DMU 3 is not very far from the least cost input mix at C.

The input overall efficiency of DMU 3 is 0.66 which is the product of its allocative cost and technical efficiency or $0.967 \times 0.682 = 0.66$. (The input overall efficiency can also be computed as follows. The aggregate cost of the input levels of DMU 3 is $2 \times \$(6 + 4) = \$20$. The least cost of producing a unit of output (point C in Figure 2.4) is $13.20. Hence the input overall efficiency of DMU 3 is 13.20 / 20 = 0.66 as obtained earlier.)

2.6. QUESTIONS

1.   Select one organisation in the public sector and one in the private sector in which you can identify homogeneous operating DMUs. Identify a set of input and output variables which can be used to assess each set of DMUs. Justify your choice of input-output variables and comment on any shortcomings they have.

2.   We have a set of DMUs which use two inputs to secure three outputs. It has been established that the maximum output levels y(50, 20, 30) a DMU can obtain from a set of input levels x(4, 5) is given by $y = \sqrt{x}$.

   i) DMU A has input levels (16, 20) and output levels (75, 30, 45). Determine its technical input and output efficiencies.
   ii) DMU B has input levels of (12, 15). What output levels would render it Pareto-efficient while maintaining its input mix?
   iii) The price per unit of input 1 is $20 and of input 2 $3. DMU A above can secure its output levels using resource levels (X1, X2) such that $X1X2 \geq 101.25$, $X1 \geq 4$, $X2 \geq 5$. What is its input allocative and what its input overall efficiency?

APPENDIX 2.1: MATHEMATICAL DEFINITIONS

Definition DA2.1.1: Pareto Efficiency

Let y_{rj} ($r = 1...s$) be the output levels secured by DMU j and x_{ij} the levels of inputs ($i = 1...m$) it uses.

Output orientation:	DMU j_0 is *Pareto-efficient* if there exists no observed or feasible in principle DMU j $\neq j_0$ such that $y_{r'j} > y_{r'j_0}$ for some r' and $y_{rj} \geq y_{rj_0}$ $\forall r \neq r'$ while $x_{ij} \leq x_{ij_0}$ $\forall i$.
Input orientation:	DMU j_0 is *Pareto-efficient* if there exists no observed or feasible in principle DMU j $\neq j_0$ such that $x_{i'j} < x_{i'j_0}$ for some i' and $x_{ij} \leq x_{ij_0}$ $\forall i \neq i'$ while $y_{rj} \geq y_{rj_0}$ $\forall r$.

Definition DA2.1.2: Technical Output Efficiency

Consider the set of DMUs referred to in DA2.1.1 above. Let $L(\underline{x})$ be the *output set* consisting of the output vectors $\underline{y}$ which can be secured by the input vector $\underline{x}$, or

$$L(\underline{x}) = \{\underline{y}: \underline{y} \text{ can be secured using inputs } \underline{x}\} \qquad (A2.1.1).$$

Let us assume that $L(\underline{x})$ is closed and convex. The technical output efficiency *TOE* of a DMU having output-input values $(\underline{y}, \underline{x})$ is $1/\theta^*$ where

$$\theta^* = \underset{\theta}{\text{Max}} \{\theta: (\theta \underline{y}) \in L(\underline{x}), \theta > 0\} \qquad (A2.1.2).$$

Definition DA2.1.3: Technical Input Efficiency

Consider the set of DMUs referred to in DA2.1.1. Let $L(\underline{y})$ be the *input requirement set* consisting of the input vectors $\underline{x}$ which can secure the output vector $\underline{y}$, or

$$L(\underline{y}) = \{\underline{x}: \underline{x} \text{ can produce } \underline{y}\} \qquad (A2.1.3).$$

Let us assume that $L(\underline{y})$ is closed and convex. The technical input efficiency TIE of $(\underline{y}, \underline{x})$ is $1/\theta^*$ where

$$\theta^* = \underset{\theta}{Max} \ \{\theta: (\underline{x}/\theta) \in L\ (\underline{y}), \theta > 0\} \tag{A2.1.4}.$$

Definition DA2.1.4: Input Overall Efficiency and Input Allocative Efficiency

Consider the set of DMUs referred to in DA2.1.1. Let input i be available to DMU k at price w_{ki} $(i = 1...m)$. The *input overall efficiency* of DMU k is

$$IOE_k = C_k\ (\underline{y}_k, \underline{w}_k)/\ \underline{w}_k\underline{x}_k \tag{A2.1.5}$$

where $\underline{w}_k = (\ w_{ki},\ i = 1...m\ )$, $\underline{x}_k = (x_{ik},\ i = 1...m)^T$. Thus $\underline{w}_k\underline{x}_k$ is the aggregate cost of the observed input levels of DMU k. $C_k\ (\underline{y}_k, \underline{w}_k)$ is the least cost at which DMU k could have secured its outputs or

$$C_k\ (\underline{y}_k, \underline{w}_k) = \underset{x_i}{Min}\ \{ \sum_i^m w_{ki}\ x_i : \underline{x} \in L\ (\underline{y_k})\} \tag{A2.1.6}$$

where $L\ (y_k)$ is as defined in (A2.1.3).

The *input allocative efficiency* of DMU k is $IAE_k = IOE_k\ /\ TIE_k$, where TIE_k is the technical input efficiency of DMU k as defined above using (A2.1.4).

APPENDIX 2.2: DERIVING GRAPHICAL MEASURES OF INPUT EFFICIENCIES

Technical input efficiency

Figure A2.1.1 is a reproduced version of Figure 2.3. The input levels along the radial line OA have the same input mix as DMU A. Moving on this line from A towards the origin the input levels reduce while the input mix of DMU A is maintained. Point G offers the lowest input levels which can secure a unit of output given the input mix of DMU A. Thus the technical input efficiency of DMU A is measured relative to point G. The level of input 1 at A can be reduced to the fraction $\dfrac{OI}{OK}$ relative to that at G and that of input 2 to the fraction $\dfrac{IG}{KA}$. The triangles OIG and OKA are similar and thus we have $\dfrac{OI}{OK} = \dfrac{IG}{KA} = \dfrac{OG}{OA}$.

Thus the technical input efficiency of DMU A is $\dfrac{OG}{OA}$.

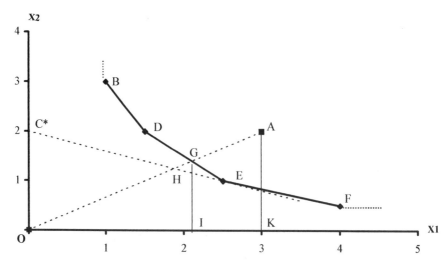

Figure A2.1.1. Deriving the Technical Input Efficiency of DMU A

Input allocative efficiency

From the derivation of the technical input efficiency above we can deduce that the levels of X1 and X2 at G are the fraction $\dfrac{OG}{OA}$ of their levels at A. By extension the levels of X1 and X2 at H, denoted respectively X1(H) and X2(H), are such that $X1(H) = \dfrac{OH}{OG} X1(G)$ and $X2(H) = \dfrac{OH}{OG} X2(G)$. Thus since at H we have cost $C^* = p_1 X1(H) + p_2 X2(H)$ at G we will have cost level

$$C(G) = p_1 X1(G) + p_2 X2(G) = \frac{OG}{OH}(p_1 X1(H) + p_2 X2(H)) = \frac{OG}{OH}C^*.$$

Thus we have:

Input allocative efficiency of A $= C^* / C(G) = \dfrac{OH}{OG}.$

Chapter 3

DATA ENVELOPMENT ANALYSIS UNDER CONSTANT RETURNS TO SCALE: BASIC PRINCIPLES

3.1 INTRODUCTION

This chapter introduces the basic principles underpinning DEA and derives special case DEA models. Chapter 4 will generalise the models introduced in this chapter.

The measures of efficiency introduced in Chapter 2 rely on estimates of maximum radial output expansion feasible for given input levels, or alternatively, on maximum radial input contraction feasible for given output levels. As we saw in Chapter 1, estimates of this kind are obtained in DEA by constructing a so-called *production possibility set* assumed to contain all input-output correspondences which are feasible in principle even if not observed in practice. The maximum feasible expansion of output levels for given input levels, or alternatively, the maximum feasible contraction of input levels for given output levels can then be estimated within the production possibility set. This process is illustrated in this chapter both graphically and by means of simple *linear programming* models. The same process is then generalised mathematically in the next chapter. As Linear Programming plays a central role in DEA a brief introduction to the subject is given in Appendix 3.1, along with references for further reading for those interested.

3.2 BASIC STEPS IN MEASURING EFFICIENCY BY DEA

The measurement of the efficiency of a DMU by DEA involves two basic steps:
- The construction of the 'Production Possibility Set' (PPS) and;
- The estimation of the maximum feasible expansion of the output or contraction of the input levels of the DMU within the PPS.

3.2.1 Constructing a Production Possibility Set in the Single-input Single-output Case

As has already been noted, the notion of the PPS is that it is a set which consists of all the input-output correspondences which are deemed feasible in principle within the input to output transformation process pertaining to the DMUs being assessed. We construct the PPS from observed input-output correspondences. To do this we make certain assumptions. Those underlying the 'basic' DEA model are as follows:

- *Interpolation* between feasible input-output correspondences leads to input-output correspondences feasible in principle;

- *Inefficient production* is possible;

- The transformation of inputs to outputs is characterised by *constant returns to scale*;

- No output is possible unless some input is used; (Informally this is known as the *'no free lunch'* assumption.)

- The PPS is *the smallest set* meeting the foregoing assumptions and containing all input-output correspondences observed.

A formal definition of the PPS and statement of the foregoing assumptions can be found in Appendix 3.2.

To illustrate the construction of a PPS we consider the simple case of the four DMUs in Table 3.1 which use a single input to secure a single output.

Table 3.1.   Observed Input-Output Correspondences

DMU	Input	Output
D1	1	2
D2	3	7
D3	4	6
D4	2	6

The input-output correspondences of Table 3.1 are labelled D1...D4 in Figure 3.1. The assumptions listed above are now used to construct a PPS.

Interpolation: Points resulting from the interpolation of two feasible input-output correspondences lie on the line joining those correspondences. For example the input-output correspondence at M is obtained by interpolation between the input-output correspondences observed at DMUs

D3 and D4. Combining 50% of the input-output levels of DMU D3 with 50% of the input-output levels of DMU D4 gives the virtual '*DMU*' at M which has input of $0.5 \times 2 + 0.5 \times 4 = 3$ and by the same token output of 6. The interpolation assumption means that we believe that 3 units of input can in principle secure 6 units of output and so some DMU could operate at M. We can make an interpolation between any number of DMUs. The resulting 'DMUs' will lie on the lines joining pairs of the original DMUs and in the space such lines enclose.

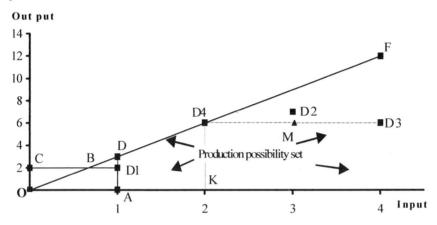

Figure 3.1. Constructing a Production Possibility Set

Inefficient production: This assumption means that feasible input-output correspondences may use input levels in excess of the minimum required for the output levels, or alternatively, may secure lower output levels than feasible for the input levels. For example by virtue of this assumption all points below the line D3D4 in Figure 3.1 and to the right of D4K are feasible. They are all less efficient than DMU D4. This assumption is formally known as *strong free disposal* of inputs and outputs as elaborated in Appendix 3.2.

Constant returns to scale (CRS): Under this assumption, if we scale the input levels of a feasible input-output correspondence up or down then another feasible input-output correspondence is obtained in which the output levels are scaled by the same factor as the input levels. Formally a definition of CRS is as follows:

Constant returns to scale:	Let $(\underline{x},\underline{y})$ be a feasible input-output correspondence where $\underline{x}$ is a set of m ≥ 1 inputs and $\underline{y}$ a set of s ≥ 1 outputs. Then $(\alpha\underline{x}, \alpha\underline{y})$ is a feasible input-output correspondence, provided $\alpha > 0$.

The implication of the CRS assumption is that the scale of operation of a DMU has no impact on its productivity. This is in contrast to other cases where we may have economies or diseconomies of scale.

In the context of Figure 3.1 the CRS assumption means that the correspondences on OD4 and its extension are feasible in principle. They are obtained by scaling up or down the feasible input-output correspondence observed at D4.

Smallest containing set: This assumption means that the PPS contains all the observed DMUs. This implicitly means that all the DMUs observed operate the same technology in the sense that they face the same production options in transforming input to output.

3.2.2 Using the PPS to Derive Efficiency Measures

Applying the assumptions listed earlier to the observed input-output correspondences of Table 3.1 we have constructed the set of feasible in principle input-output correspondences denoted *production possibility set* in Figure 3.1. The line OF is known as the *efficient boundary* of the PPS because all points on it are Pareto-efficient. That is no point on this boundary is dominated by some other feasible point which can offer higher output without requiring higher input or use lower input without lowering output also. The efficiencies of DMUs are measured using OF as reference or 'benchmark'.

We can now use the definition of technical input efficiency given in Chapter 2 to compute that of a DMU such as D1 in Figure 3.1. Using the efficient boundary OF as a benchmark we can estimate the minimum input level which could have secured the output of 2 offered by DMU D1. If we maintain the output level of D1 at 2 but lower its input level we obtain new feasible input-output correspondences on BD1. Thus the lowest input level D1 could have used is CB and so
   the technical input efficiency of D1 is CB / CD1.

In a similar manner, if we maintain the input level of DMU D1 at 1 but raise its output level we obtain new feasible input-output correspondences on D1D. Expressing the output level AD1 of DMU D1 as a proportion of the maximum feasible output AD for its input we conclude that
   the technical output efficiency of D1 is AD1 / AD.

From the similarity of the triangles COB and OAD and given that OA = CD1 and OC = AD1 we can readily deduce that CB / CD1 = AD1 / AD which means that

the technical input and output efficiencies of DMU D1 are equal.

This is as we would expect under CRS, (Cooper et al. (2000) section 3.8.)

3.3 USING LINEAR PROGRAMMING TO MEASURE EFFICIENCY IN THE SINGLE-INPUT SINGLE-OUTPUT CASE

We can measure the efficiency of a DMU such as D1 in Figure 3.1 using linear programming rather than graphical means. The use of linear programming is essential for general purpose assessments of efficiency by DEA as graphical means can only be employed where the DMUs use at most a total of three inputs and outputs. This section introduces a simple linear programming model to simulate the measurement of the efficiency of D1 effected graphically in the previous section.

The assumptions we used to construct the PPS in Figure 3.1 can be captured within a linear programming model as follows. The *interpolation* assumption means that any input-output correspondence (X, Y) is feasible in the context of the DMUs listed in Table 3.1 provided that

$$\begin{pmatrix} X \\ Y \end{pmatrix} = \Lambda_1 \begin{pmatrix} 1 \\ 2 \end{pmatrix} + \Lambda_2 \begin{pmatrix} 3 \\ 7 \end{pmatrix} + \Lambda_3 \begin{pmatrix} 4 \\ 6 \end{pmatrix} + \Lambda_4 \begin{pmatrix} 2 \\ 6 \end{pmatrix} \text{ and}$$

$$\Lambda_1 + \Lambda_2 + \Lambda_3 + \Lambda_4 = 1 \qquad \Lambda_j \geq 0, \qquad j = 1...4.$$

This reduces to

$$\begin{aligned} X &= \Lambda_1 + 3\Lambda_2 + 4\Lambda_3 + 2\Lambda_4 \\ Y &= 2\Lambda_1 + 7\Lambda_2 + 6\Lambda_3 + 6\Lambda_4 \\ \Lambda_1 + \Lambda_2 + \Lambda_3 + \Lambda_4 &= 1 \qquad \Lambda_j \geq 0, \qquad j = 1...4. \end{aligned} \tag{3.1}$$

The assumption that the *DMUs operate under CRS* makes the restriction $\Lambda_1 + \Lambda_2 + \Lambda_3 + \Lambda_4 = 1$ unnecessary, provided that at least one Λ is positive. To see this note that if we omit the restriction $\Lambda_1 + \Lambda_2 + \Lambda_3 + \Lambda_4 = 1$ from (3.1) then the derived input-output correspondence will be (x, y), where

$$\begin{aligned} x &= \lambda_1 + 3\lambda_2 + 4\lambda_3 + 2\lambda_4 \\ y &= 2\lambda_1 + 7\lambda_2 + 6\lambda_3 + 6\lambda_4 \\ \lambda_1 + \lambda_2 + \lambda_3 + \lambda_4 &= k, \qquad \lambda_j \geq 0, \qquad j = 1...4. \end{aligned} \tag{3.2}$$

Since the DMUs operate under CRS, provided k > 0, we can divide through by k both x and y in (3.2) to arrive at the expression in (3.1). We shall have in (3.1) $\Lambda_j = \lambda_j / k$, $j = 1...4$ and $\sum_{j=1}^{4} \Lambda_j = 1$ as required. k will be positive in (3.2) if at least one λ_j is positive. This is tantamount to saying that provided at least one observed DMU is used to derive feasible input-output correspondences by means of (3.2), k will be positive.

The assumption that *production could be inefficient* means that in (3.2) x may exceed and y may be below the corresponding RHS.

Summarising the foregoing steps and observations we conclude that (x,y) will be a feasible input-output correspondence within the technology operated by the DMUs of Table 3.1 (i.e. will be within the PPS) provided that:

$$\begin{aligned} x &\geq \lambda_1 + 3\lambda_2 + 4\lambda_3 + 2\lambda_4 \\ y &\leq 2\lambda_1 + 7\lambda_2 + 6\lambda_3 + 6\lambda_4 \quad\quad (3.3) \\ \lambda_j &\geq 0 \quad\quad j = 1...4. \end{aligned}$$

To measure now the technical input efficiency of D1 we need to identify a correspondence (x,y) which is feasible within (3.3) and offers the maximum contraction to the input level of D1 without detriment to its output level. This we can do by solving the linear programming model in [M3.1].

$$\begin{aligned} &\text{Min} && z && \text{[M3.1]} \\ &\text{Subject to:} && z \geq \lambda_1 + 3\lambda_2 + 4\lambda_3 + 2\lambda_4 \\ & && 2 \leq 2\lambda_1 + 7\lambda_2 + 6\lambda_3 + 6\lambda_4 \\ & && \lambda_j \geq 0 \quad\quad j = 1...4, \quad z \text{ free.} \end{aligned}$$

The constraints of [M3.1] are the expressions for the input-output correspondence (x, y) in (3.3) where x = z and y = 2. So any feasible set of λ values in [M3.1] will yield an input-output correspondence within the PPS of the DMUs in Table 3.1. Further, because we are minimising z the input level of this correspondence (which is $\lambda_1 + 3\lambda_2 + 4\lambda_3 + 2\lambda_4$) is being minimised while the output level (which is $2\lambda_1 + 7\lambda_2 + 6\lambda_3 + 6\lambda_4$.) is not permitted to drop below the 2 units of output that D1 secures.

When we solve [M3.1] (e.g. using Excel or linear programming software see Appendix 3.1) we get input level of z* = 0.6667 = $\lambda_1^* + 3\lambda_2^* + 4\lambda_3^* + 2\lambda_4^*$ and output level of $2\lambda_1^* + 7\lambda_2^* + 6\lambda_3^* + 6\lambda_4^* = 2$. (The * denotes the optimal value of the corresponding variable.) Thus by solving [M3.1] we have identified a feasible 'virtual DMU' which will deliver the output of D1 (2 units) using a minimum feasible input of 0.6667 units. By definition therefore the technical input efficiency of D1 is 0.6667 / 1 or 66.67%.

The solution to the model yields further information. All λ's are zero except λ_4^* which is 0.3333. If we plug this information in the RHSs of [M3.1] we will deduce that if we scale D4 down to 33.33% of its original levels we will get input of $2 \times 0.3333 = 0.6666$ and output $6 \times 0.3333 = 2$ units. This scaled down version of D4 therefore has been used by the model as a *benchmark* in estimating the technical input efficiency of D1. As we will see later identification of benchmarks of this kind has great practical importance within the broader framework of performance management.

Model [M3.1] can be readily modified to assess the technical input efficiency of some other DMU in Table 3.1. For example to assess the technical input efficiency of D2 model [M3.1] becomes as in [M3.2].

Min   z   [M3.2]
Subject to:

$$3z \quad \geq \lambda_1 + 3\lambda_2 + 4\lambda_3 + 2\lambda_4$$
$$7 \quad \leq 2\lambda_1 + 7\lambda_2 + 6\lambda_3 + 6\lambda_4$$
$$\lambda_j \quad \geq 0 \qquad j = 1...4, \quad z \text{ free.}$$

The RHSs of models [M3.1] and [M3.2] are identical as both DMUs operate within the same PPS. The LHSs of [M3.1] and [M3.2] differ, each one reflecting the observed input-output correspondence at the DMU being assessed. At the optimal solution of [M3.2] we get input level of $3z^* = 2.3334 = \lambda_1^* + 3\lambda_2^* + 4\lambda_3^* + 2\lambda_4^*$ and output level of $2\lambda_1^* + 7\lambda_2^* + 6\lambda_3^* + 6\lambda_4^* = 7$. Thus the technical input efficiency of D2 is 2.334 / 3 = 0.7778 or 77.78%. This efficiency rating is of course the optimal value of z, $z^* = 0.7778$.

At the optimal solution of [M3.2] only λ_4 is positive at the value of 1.167 and all other λ's are zero. This means that in the case of D2 as in that of D1 the 'benchmark' is again D4 only this time scaled up by the factor of 1.167. We can readily see that at this scaled up version D4 would yield the output level of $6 \times 1.167 = 7$ using input of $2 \times 1.167 = 2.334$ as estimated above for D2.

Finally, the model in [M3.1] can be readily modified to assess the technical output rather than input efficiency of a DMU. For example to assess the technical output efficiency of DMU D1 model [M3.1] is modified to that in [M3.3].

Max   h   [M3.3]
Subject to:

$$1 \quad \geq \alpha_1 + 3\alpha_2 + 4\alpha_3 + 2\alpha_4$$
$$2h \quad \leq 2\alpha_1 + 7\alpha_2 + 6\alpha_3 + 6\alpha_4$$
$$\alpha_j \quad \geq 0 \qquad j = 1...4, \quad h \text{ free.}$$

The model in [M3.3] identifies non-negative α^* values which lead to a point within the PPS which offers maximum output level $2\alpha_1^* + 7\alpha_2^* + 6\alpha_3^* + 6\alpha_4^* = 2h^*$ for given input level $\alpha_1^* + 3\alpha_2^* + 4\alpha_3^* + 2\alpha_4^* = 1$. (The * is used to denote the optimal value of the corresponding variable.) <u>Thus h* is the maximum factor</u> by which the output of D1 can be expanded without detriment to its input level. Following the definition of technical output efficiency in Chapter 2 we deduce that the technical output efficiency of D1 will be $\frac{1}{h^*}$.

When we solve [M3.3] we get $h^* = 1.5$. Thus the technical output efficiency of D1 is $1/1.5 = 0.6666$ or 66.66%. This means that its observed output level of 2 is only 66.66% of the maximum output level D1 could secure for its input level. The technical output efficiency of D1 thus matches its technical input efficiency at 66.66%. This is not coincidental.

It can be shown (e.g. see Cooper et al. (2000) section 3.8) that technical input and output efficiencies are equal under CRS.

At the optimal solution to [M3.3] we have $\alpha_4^* = 0.5$ and all other α's are zero. This means that in the output as in the input oriented measure of the efficiency of D1 the 'benchmark' unit is D4, only this time scaled down by the factor of 0.5. We can readily see that D4 scaled to half its size would yield the output level of 3 using input of 1 which matches that of D1. However, the output level of D1 is only two-thirds (i.e. 66.66%) of that of the scaled down D4 and this underlies the technical output efficiency rating model [M3.3] yielded in respect of D1.

The single-input single-output example used so far to illustrate the measurement of technical input and output efficiency in DEA does not lend itself for demonstrating certain features which arise in assessing relative efficiency in the more general multi-input multi-output case. Such features are demonstrated in the next three sections using multi-input or multi-output examples but not both.

3.4 USING DEA TO MEASURE TECHNICAL INPUT EFFICIENCY IN THE SINGLE-OUTPUT MULTI-INPUT CASE: A GRAPHICAL ILLUSTRATION

To illustrate what is involved here we use a two-input one-output example. We wish to assess the technical input efficiencies of four tax offices. We shall assume that the size of the resident population a tax office serves is a good proxy for its volume of activities and so we use *resident population served* as our single output. We shall assume that the tax offices use two inputs to serve the resident population: *Operating expenditure* (OPEX) and *capital expenditure* (CAPEX). OPEX covers all recurrent expenditure in running the tax offices (including staff costs) while CAPEX covers capital expenditure (including computing equipment, investment in Information Technology and so on). (The specification of the input-output set is of paramount importance in any DEA-based assessment of efficiency as noted earlier. However, in order to concentrate on the principles underlying DEA measures of efficiency, we shall at this stage avoid the complexities of identifying a more appropriate input-output set for assessing tax offices. We shall also ignore at this stage the difficulties of measuring CAPEX which, unlike OPEX, impacts performance over many time periods.)

We assume that the operations of tax offices are characterised by CRS. Table 3.2 shows the OPEX and CAPEX levels per resident. (The data is fictitious.)

Table 3.2.   Input Levels at Tax Offices ($ per Resident)

DMU	OPEX	CAPEX
TO1	1	3
TO2	3	2
TO3	3.75	1
TO4	1.5	1.5

As we have standardised the output, the offices differ only in the input levels they use per unit of output. Since we have only two inputs we can plot the offices in two-dimensions as shown in Figure 3.2.

Using now the assumptions we employed in respect of Figure 3.1 we can construct the PPS relating to the DMUs in Table 3.2. For example we may interpolate between tax offices TO1 and TO2 to construct a feasible 'tax office' mid-way between them at point M, Figure 3.2. Moreover, using the assumption that inefficient transformation of inputs to outputs is possible then if a tax office can operate at M so can tax offices in the shaded area to the right and above M which are dominated by M. Extending the approach followed in respect of Figure 3.1 to the four tax offices in Table 3.2 leads to the PPS shown in Figure 3.3. The CAPEX-OPEX levels on, to the right and above the lines TO1TO4TO3 and incorporating all observed tax offices constitute the PPS. All points within the PPS can secure a unit of output.

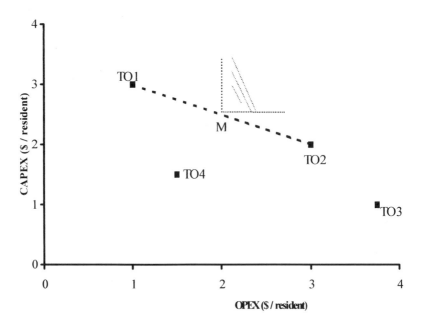

Figure 3.2.   Tax Offices Plotted in Input Space

The piece-wise linear boundary TO1TO4TO3 to the PPS in Figure 3.3 is the locus of the Pareto-efficient points of the PPS. The points on this boundary are used as referents for measuring the efficiencies of the points within the PPS.

We have defined the technical input efficiency of a tax office as the proportion to which its input levels can be contracted radially (i.e. keeping its input mix constant), without detriment to its output level. We will now see how we can compute this efficiency measure for TO2 using Figure 3.4 which contains the PPS we constructed in Figure 3.3.

The input mix of TO2 defines the radial line ATO2 in Figure 3.4. To estimate the proportion to which the input levels of tax office TO2 can be lowered while maintaining its input mix we must estimate how far point TO2 can be moved towards the origin along the radial line ATO2. The point *b'* where ATO2 crosses the boundary TO1TO4TO3 offers the lowest absolute levels of input of the same mix as those of tax office TO2, which are capable of securing a unit of output. Thus point *b'* becomes the *referent* point relative to which the efficiency of tax office TO2 is measured. The technical input efficiency of tax office TO2 is the proportion the input levels of *b'* represent of those of TO2, as both TO2 and *b'* secure the same output level.

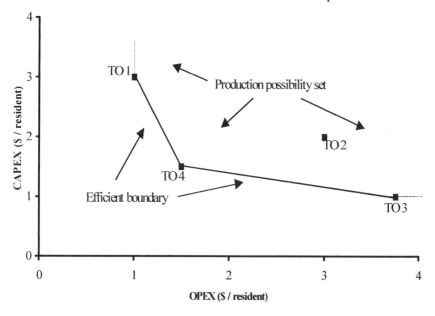

Figure 3.3.   The Production Possibility Set (PPS) Derived From TO1-TO4

Following the procedure outlined in Appendix A2.2 (in Chapter 2) we can deduce that the input levels at *b'* represent a proportion *Ab' / ATO2* of those of tax office TO2.

Thus the technical input efficiency of tax office TO2 is Ab' / ATO2.

Note that the definition of technical input efficiency in Chapter 2 required that the outputs not deteriorate rather than that they stay constant as happens implicitly above in Figure 3.4. Nevertheless our approach has led to a correct measure of the efficiency of TO2 because in estimating the lowest levels its inputs could take without detriment to its output level it would not make sense to raise its output level. (The truth of this intuitively obvious statement is shown mathematically in the next section.)

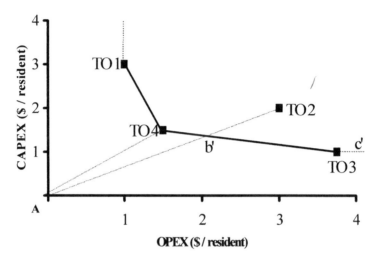

Figure 3.4. Measuring Technical Input Efficiency

It is relatively easy to see that any tax office which is located on the boundary of the PPS will have technical input efficiency of 1. For example the reference point for measuring the technical input efficiency of tax office TO4 is point TO4 itself. This is the only point common to the radial ATO4 and the PPS. Thus the technical input efficiency of tax office TO4 is $ATO4 / ATO4 = 1$.

As we saw in Chapter 2 we can have boundary points which are not Pareto-efficient. A case in point would be some DMU located say at c′ in Figure 3.4. Point c′ is not Pareto-efficient as it is dominated by DMU TO3. Both deliver the same output and use the same level of input 2 but TO3 uses a lower level of input 1. It can similarly be seen that all points on the horizontal line from TO3 and the vertical line from TO1 are Pareto-inefficient boundary points. Yet these points have technical input efficiency of 1 because it is impossible to lower simultaneously both their input levels while maintaining their input mix.

In summary, all PPS boundary points have technical input efficiency of 1 but only some of them are Pareto-efficient.

Note that the DEA measure of efficiency obtained for each tax office is *relative* to the PPS used which in turn is determined by the observations used to construct it. If further tax offices are added to those used to construct the PPS in Figure 3.3 (and 3.4) the PPS could alter and so can the efficiencies of the tax offices. This is true of DEA in general in that the measure of efficiency of a DMU can change when the set of DMUs used to construct the PPS changes. Further, a DMU may well be inefficient in absolute terms even when efficient in DEA terms. **All that the DEA efficiency measure tells us is whether or not a DMU can improve its performance relative to the set of DMUs to which it is being compared.**

3.5 USING LINEAR PROGRAMMING TO MEASURE TECHNICAL INPUT EFFICIENCY IN THE SINGLE-OUTPUT MULTI-INPUT CASE

As in the single-input single-output case so too in the multi-input single-output case depicted in Figure 3.4 we can measure technical input efficiency using a linear programming model. Moreover, the same model can also be used to identify whether a point is Pareto-efficient or not.

Using the assumptions of interpolation, CRS and the possibility of inefficient production we derive that with reference to the DMUs in Table 3.2 any set (x_1, x_2, y) of OPEX, CAPEX and Resident levels respectively which meets the following restrictions is feasible:

$$x_1 \geq \lambda_1 + 3\lambda_2 + 3.75\lambda_3 + 1.5\lambda_4 : \quad :OPEX$$

$$x_2 \geq 3\lambda_1 + 2\lambda_2 + \lambda_3 + 1.5\lambda_4 \quad :CAPEX$$

$$y \leq \sum_{j=1}^{4} \lambda_j \quad :RESIDENTS^\$ \quad (3.4)$$

$$\lambda_j \geq 0 \quad j = 1...4.$$

$ It is recalled that in Table 3.2 the output of all tax offices was standardised to 1 resident.

The values of (x_1, x_2, y) defined in (3.4) constitute the PPS which can be constructed using the four tax offices in Table 3.2. The interpolation assumption requires the additional constraint $\sum_{j=1}^{4} \lambda_j = 1$. This is, however, unnecessary under CRS as we saw in arriving at (3.3) earlier.

In order to now determine whether a tax office is Pareto-efficient and to estimate its technical input efficiency we need to identify a point within the PPS which offers the maximum radial contraction of the input levels of the tax office without detriment to its output level. The following linear programming model fulfils this objective in the case of tax office TO2 in Table 3.2:

Min $\quad$ q - ε (S_OPEX + S_CAPEX + S_ RESP) $\hspace{2cm}$ [M3.4]
Subject to:

$$3\,q - S\_OPEX = \lambda_1 + 3\lambda_2 + 3.75\lambda_3 + 1.5\lambda_4 \quad :OPEX$$

$$2\,q - S\_CAPEX = 3\lambda_1 + 2\lambda_2 + \lambda_3 + 1.5\lambda_4 \quad :CAPEX$$

$$1 + S\_RESP = \sum_{j=1}^{4} \lambda_j \hspace{3cm} :RESIDENTS$$

$\lambda_1 \dots \lambda_4,$ S_OPEX, S_CAPEX, S_ RESP ≥ 0, q free,

$0 < ε \ll 1$ is a *non-Archimedean infinitesimal.*

The non-Archimedian infinitesimal ε is meant to be 'smaller than any positive number'. This is only a mathematical construct to ensure we identify Pareto-efficient points as will be explained below. **In practice we do <u>not</u> use a real number for ε** but rather q is minimised first and then the sum of slacks is maximised keeping q to its minimum value obtained.

Let the superscript * denote the optimal value of the corresponding variable in [M3.4]. Within the structure of [M3.4] q represents the radial contraction to the input levels of tax office TO2. As noted above we solve [M3.4] by first minimising q.

The model in [M3.4] will always yield q* ≤ 1 as by setting $\lambda_2 = 1$ and all other variables to zero we would obtain a feasible solution in which q = 1. Further, it is easy to see that we cannot have S_ RESP* > 0 as we could always lower the λ values to drive S_ RESP* to zero, lowering at the same time the value of q which accords with the objective function of the model. It is, however, possible to have S_OPEX* > 0 or S_CAPEX* > 0 (but not both as again we could lower q to drive at least one of these two slack variables to zero). We will now interpret the solution to [M3.4] for the two cases, one where the optimal value of q* is 1 and the other where q* is below 1.

Case where q*=1

Two possibilities arise here. One is the case where none of the slack values is positive (i.e. S_OPEX* = S_CAPEX* = S_ RESP* = 0). The other is where one of the slack values is positive.

The case where q* = 1 and S_OPEX* = S_CAPEX* = S_ RESP* = 0 is saying that the input levels of tax office TO2 cannot be contracted jointly or individually without detriment to its output level and neither can that output level be raised, given its input levels. This would make **tax office TO2 Pareto-efficient, with technical input efficiency of 1**.

When q* = 1 and one slack value is positive at the optimal solution to [M3.4] we have identified a PPS point which offers the output level of tax office TO2 using the same input level as TO2 on one input, but less of the input with the positive slack. This makes **tax office TO2 not Pareto-efficient but with radial efficiency of 1** as it is not possible to contract both its inputs.

Case where q*<1

In this case model [M3.4] will have identified a PPS point which offers the output level of tax office TO2 using a proportion q* < 1 of its input levels. Thus in this case **tax office TO2 is not Pareto efficient.** Further, **q* measures its technical input efficiency as it represents the maximum feasible *radial* contraction to its inputs without detriment to its output level**. If not only we have q* < 1 but also some slack is positive at the optimal solution to [M3.4], then the input corresponding to the positive slack can be lowered below the proportion q* of its original level, without detriment to TO2's output level.

Model [M3.4] can be readily adapted to ascertain the technical input efficiency of any one of the tax offices in Table 3.2. It is merely necessary to alter the coefficients of q to reflect the OPEX and CAPEX levels of the tax office being assessed. It would also normally be necessary to adjust the LHS of the output constraint but this is not necessary here because output has been standardised to 1 resident across all tax offices. Further, it is relatively easy to see that had there been additional inputs to CAPEX and OPEX model [M3.4] would have been augmented by constraints relating to the additional inputs. Such constraints would have had the structure of the constraints labelled OPEX and CAPEX. The same goes for additional outputs which would have generated within [M3.4] constraints such as that relating to resident population. The formal generalisation of [M3.4] to the multi-input-multi-output case is given in Chapter 4.

In conclusion, model [M3.4], with suitable coefficients of q can be used to assess any one of the tax offices in Table 3.2. Using a superscript * to denote the optimal value of a variable we deduce that the tax office which is the subject of the model:

– Is Pareto-efficient if and only if $q^* = 1$ and $S\_OPEX^* = S\_CAPEX^* = S\_RESP^* = 0$;
– Has technical input efficiency of q^*.

3.6   USING DEA TO MEASURE TECHNICAL OUTPUT EFFICIENCY IN THE SINGLE-INPUT MULTI-OUTPUT CASE: A GRAPHICAL ILLUSTRATION

Technical output efficiency in the single-input multi-output case can be measured using DEA in a similar manner as measuring technical input

Table 3.3.   Operating Expenditure and Activity Levels

Hospital	Operating Expenditure ($000)	Number of severe patients	Number of regular patients
A	300	900	2100
B	250	1500	1500
C	320	2560	1280
D	400	3200	800
E	100	500	400
F	80	480	160

efficiency in the single-output multi-input case covered in the previous two sections. To illustrate the approach consider a Regional Health Authority which manages 6 hospitals. The data on operating expenditure and activity volumes for the latest full accounting year are shown in Table 3.3 (source Dyson et al. 1993). Let us use DEA to assess the technical output efficiency of hospital E.

Clearly operating expenditure is the input while regular and severe patient levels are the outputs here. Let us assume that the operations of the hospitals are characterised by CRS. Then we can standardise the input to say $1000 and the corresponding output levels appear in Table 3.4.

Table 3.4.   Patients per $1000 Operating Expenditure

Hospital	Severe	Regular
A	3	7
B	6	6
C	8	4
D	8	2
E	5	4
F	6	2

Using the assumptions deployed with reference to Figures 3.1 and 3.3 we can construct the PPS depicted in Figure 3.5. The PPS is bounded from above by

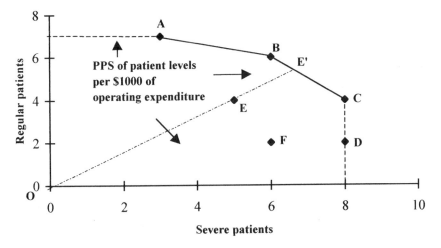

Figure 3.5.   Computing the Technical Output Efficiency of Hospital E

the piece-wise linear boundary ABC and by the broken horizontal line from A and the broken vertical line from C. The Pareto-efficient output levels lie on ABC while the output levels on the broken parts of the boundary are not Pareto-efficient as they are dominated by the output levels at A (horizontal broken line) and C (vertical broken line).

To compute the technical output efficiency of hospital E we need to estimate the factor by which its output levels can be expanded radially (i.e. while maintaining its output mix) while keeping its input level at $1000. The output mix of hospital E defines the radial OEE′ in Figure 3.5. The output levels at E′ are maximum within the PPS for the output mix of hospital E. Thus E′ becomes the referent point within the PPS for measuring the technical output efficiency of hospital E. Following the derivation in Appendix 2.2 of Chapter 2 we conclude that

the output efficiency of hospital E is $\dfrac{OE}{OE'}$.

3.7 USING LINEAR PROGRAMMING TO MEASURE TECHNICAL OUTPUT EFFICIENCY IN THE SINGLE-INPUT MULTI-OUTPUT CASE

We can use linear programming to measure the technical output efficiency of a DMU and identify whether it is Pareto-efficient in the multi-output case much as we did in the multi-input case. The model is introduced using the hospitals in Table 3.3.

Using the assumptions of interpolation, CRS and the possibility of inefficient production we derive that any set (y_1, y_2, x) of severe patient, regular patient and operating expenditure levels respectively which meets the following restrictions is feasible:

Severe patients:

$$y_1 \geq 900\lambda_A + (1500 + \lambda_B) + 2560\lambda_C + 3200\lambda_D + 500\lambda_E + 480\lambda_F$$

Regular patients:

$$y_2 \geq 2100\lambda_A + 1500\lambda_B + 1280\lambda_C + 800\lambda_D + 400\lambda_E + 160\lambda_F \qquad (3.5)$$

OPEX:

$$x \leq 300\lambda_A + 250\lambda_B + 320\lambda_C + 400\lambda_D + 100\lambda_E + 80\lambda_F$$
$$\lambda_j \geq 0 \qquad j = A...F.$$

(The interpolation assumption requires additionally the constraint $\sum_{j=A}^{F} \lambda_j = 1$. This, however, need not be included under the CRS assumption as we saw in arriving at (3.3) earlier.) In order to determine whether a hospital is Pareto-efficient and estimate its technical output efficiency we need to identify some point within the PPS which offers the maximum radial expansion of the hospital's output levels without detriment to its input level. The following linear programming model fulfils this objective in the case of Hospital E:

Max $\qquad$ $h + \varepsilon(S\_SEVERP + S\_REGP + S\_OPEX)$ $\qquad$ [M3.5]

Subject to:

Severe patients: $500\,h + S\_SEVERP =$
$$900\lambda_A + 1500\lambda_B + 2560\,\lambda_C + 3200\lambda_D + 500\lambda_E + 480\,\lambda_F$$

Regular patients: $400\,h + S\_REGP =$
$$2100\lambda_A + 1500\lambda_B + 1280\lambda_C + 800\lambda_D + 400\lambda_E + 160\lambda_F$$

Operating expenditure: $100 - S\_OPEX =$
$$300\lambda_A + 250\lambda_B + 320\lambda_C + 400\lambda_D + 100\lambda_E + 80\lambda_F$$

$\lambda_j \geq 0, \qquad j = A...F,$

$S\_OPEX, S\_REGP, S\_SEVERP \geq 0$, h free,

$0 < \varepsilon \ll 1$ is a non-Archimedean infinitesimal.

Let the superscript * denote the optimal value of the corresponding variable in [M3.5]. As customary we use the non-Archimedean ε in model [M3.5] to indicate that we give pre-emptive priority to the maximisation of h which represents the radial expansion factor of the output levels of hospital E. The model will always yield $h^* \geq 1$ as by setting $\lambda_E = 1$ and all other variables to zero we would obtain a feasible solution in which $h = 1$.

If we obtain $h^* = 1$ and $S\_OPEX^* = S\_REGP^* = S\_SEVERP^* = 0$ then we would have determined by means of [M3.5] that the output levels of hospital E cannot be expanded jointly or individually without raising its input level. Further, its input level cannot be lowered given its output levels. This would make hospital E **Pareto-efficient with technical output efficiency of 1.**

If we obtain $h^* = 1$ but one slack value is positive at the optimal solution to [M3.5] then the model would have identified a PPS point which uses the input level of hospital E and offers the same level on one of the outputs as hospital E, but it offer in excess of hospital E on the output corresponding to the positive slack. This would make hospital E **not Pareto-efficient but with radial efficiency of 1** as its outputs cannot be expanded jointly. (At the optimal solution to model [M3.5] we can have no more than one of the two output related slacks positive for the same reason as we could have at most one input slack positive at the optimal solution to model [M3.4].)

Finally we could have at the optimal solution to [M3.5] $h^* > 1$. The model will have identified in this case a point within the PPS which offers regular and severe patient levels in excess of those of hospital E for no additional input. Thus hospital E would not be Pareto-efficient in this case. h^* is the maximum factor by which <u>both</u> its observed output levels can be expanded without the need to raise its input. Thus, (see Chapter 2)

the technical output efficiency of hospital E is (1 / h*) < 1.

If at the optimal solution to [M3.5] we have not only h* > 1 but also some slack positive, then the output of hospital E corresponding to the positive slack can be raised by more than the factor h*, without the need for additional input. This potential additional output at hospital E is not reflected in its efficiency measure because the additional output does not apply across all output dimensions.

Model [M3.5] can be readily adapted to ascertain the technical output efficiency of any one of the hospitals in Table 3.3. It is merely necessary to alter the coefficients of h and the LHS to the operating expenditure constraint to reflect the input-output levels of the hospital being assessed. Further, it is relatively easy to see that had there been additional outputs to severe and regular patients model [M3.5] would have been augmented by constraints relating to the additional outputs. The same goes for additional inputs which would have generated within [M3.5] constraints such as that relating to operating expenditure. The formal generalisation of [M3.5] to the multi-input multi-output case is given in Chapter 4.

In conclusion, having solved model [M3.5] with respect to a hospital we deduce:

- The hospital is Pareto-efficient if and only if h* = 1 and S_OPEX* = S_REGP* = S_SEVERP* = 0;
- The hospital has technical output efficiency 1/h*.

The next chapter generalises the DEA models presented in this chapter.

3.8   QUESTIONS

1.   A bank has five branches. The table below shows the number of hours of staff time used weekly per 1000 accounts in existence at the branch.

Table 3.5.   Labour Hours per 1000 Accounts

BRANCH	Supervisory	Trainee
1	2	3
2	4	1
3	2	2
4	1	4
5	1	5

a) Construct a graphical representation of the PPS of supervisory and trainee hours capable of handling 1000 accounts.

b) Using the PPS constructed compute the technical input efficiency of Branch 1.

2.   Formulate and solve a linear programming model to ascertain whether Branch 4 of question 1 above is Pareto-efficient and compute its technical input efficiency. Repeat for Branch 5 and comment on the results for Branches 4 and 5.

3.   Solve model [M3.4] to obtain the technical input efficiency of tax office 2 of Table 3.2. Formulate and solve a linear programming model to ascertain the technical output efficiency of the same tax office. Comment on the efficiency rating of the tax office by the two alternative models.

4.   In a region of the country there are five local tax offices (Table 3.6).
        After some consideration it is decided that the volume of work of the
        offices can be reflected by the following:

-Number of accounts administered;

-Number of tax adjustment applications processed;

-Number of summonses and distress warrants issued;

-Net Present Value (NPV) of taxes collected.
The resources used in processing the work of a tax office are reflected
in its operating expenditure (OPEX). Data for the most recent
financial year are as follows:

Table 3.6.   Data on a Set of Tax Offices

Office	Running Cost	No of accounts	No of adj. Appl'ns	No of Sum-monses	NPV of Taxes
	($00,000)	(0,000)	(000)	(000)	($10m)
1	9.13	7.52	34.11	21.96	3.84
2	13.60	8.30	23.27	46.00	8.63
3	5.65	1.78	0.16	1.31	39.01
4	12.71	6.32	13.63	8.53	5.16
5	11.19	6.58	10.90	3.52	3.46

In respect of office 4:

i.   Set up the DEA model to assess its technical input efficiency.
ii.   Determine its technical input efficiency.
iii.   Set up the DEA model to assess its technical output
        efficiency.
iv.   Determine its technical output efficiency.
v.   Contrast the efficiency estimates obtained by the alternative
        DEA models used.

APPENDIX 3.1: INTRODUCTION TO LINEAR PROGRAMMING

It is beyond the scope of this book to give full coverage of linear programming whose use, of course, extends far beyond its specialist use in the framework of DEA. The interested reader can find more thorough coverage of linear programming in a variety of textbooks, including Winston (1994) and Anderson et al. (1999). This Appendix gives a brief introduction which will enable the reader to understand in a very basic way the linear programming models presented in this book within the context of DEA.

Formulation of Linear Programming Models

Linear Programming is a method of mathematical analysis that seeks the best (*optimal*) solution within an infinite set of available solutions, the solutions typically referring to alternative available courses of action in some decision making context. The process begins with the construction of a mathematical model the "*linear programming model*" which mirrors the decision situation being addressed. That is the model defines mathematically all potential or *feasible* solutions and also contains a mathematical expression which is used to evaluate the attractiveness of each feasible solution. An example will illustrate how we build such models.

Example A3.1.1: A firm produces three products which require grinding, polishing and packing. The number of hours each product requires on each process and the times available are as follows:

Table 3.7. Processing Time (hours) per Unit of Each Product

Process	Product 1	Product 2	Product 3	Available hours per week (hours)
Grinding	3	6	8	700
Polishing	3	6	4	480
Packing	4	3	2	650

Products 1, 2 and 3 sell at $35, $20 and $15 per unit respectively. Formulate a Linear Programming model to determine the units of each product to be made weekly to maximise sales revenue.

Solution: Let $PROD_1$, $PROD_2$ and $PROD_3$ be respectively the units of product 1, 2 and 3 to be made weekly. These are referred to as *decision variables*. The ultimate aim of the analysis is to find out the exact values they should take to maximise total weekly sales revenue.

Given that 700 hours are available weekly for grinding and given the grinding time required per unit of each product the values of the decision variables must satisfy the following constraint:

Grinding: $3PROD_1 + 6PROD_2 + 8PROD_3 \leq 700$.

In a similar manner the limited weekly hours available for polishing and packing give rise to the following constraints.

Polishing: $3PROD_1 + 6PROD_2 + 4PROD_3 \leq 480$
Packing: $4PROD_1 + 3PROD_2 + 2PROD_3 \leq 650$.

It would not make sense for any of $PROD_1$, $PROD_2$ and $PROD_3$ to be negative. Hence we need the restrictions

$PROD_1$, $PROD_2$, $PROD_3 \geq 0$.

Any set of values for $PROD_1$, $PROD_2$ and $PROD_3$ that meet the foregoing constraints and non-negativity restrictions represent a *feasible* production plan. We need to choose the best feasible production plan, namely the one that maximises weekly sales revenue.

The weekly sales revenue in ($) is: $35PROD_1 + 20PROD_2 + 15PROD_3$.

This is referred to as the *objective function*. It is the function we wish to *optimise*.

So, we have formulated the following Linear Programming model:

Maximise $35PROD_1 + 20PROD_2 + 15PROD_3$ [MA3.1.1]
Subject to:

$3PROD_1 + 6PROD_2 + 8PROD_3 \leq 700$
$3PROD_1 + 6PROD_2 + 4PROD_3 \leq 480$
$4PROD_1 + 3PROD_2 + 2PROD_3 \leq 650$
$PROD_1$, $PROD_2$, $PROD_3 \geq 0$.

The solution of this model will give the production plan that maximises weekly sales revenue.

Generalising from the foregoing example we note that a linear programming model has the following features:

- A set of *decision variables* which stand for unknown quantities;
- A set of *constraints* which define feasible values of the decision variables;
- An *objective function* which reflects some pay-off such as profit or cost associated with each feasible set of values of the decision variables.

The solution of the model leads to the identification of a feasible set of values of the decision variables which optimise the objective function.

Solving Linear Programming Models

Linear Programming models can be solved using a variety of methods, most notably the *Simplex Method*. The interested reader can find details of this method in a variety of textbooks on linear programming, including Winston (1994) and Anderson et al. (1999). The methods are implemented in computer software. The most readily available software for this purpose is probably in Excel, under *Solver* in the *Tools* menu. Examples of specialist software which include a linear programming facility are 'XPRESSMP' and 'LINDO'. Free demonstration versions of these software are available on the internet at the time of writing. LINDO can be downloaded from http://www.lindo.com and a student version of Visual XPRESS (i.e. Windows version) can be downloaded from http://www.dash.co.uk/. (All other XPRESS software is available (at the time of writing) free to users as a loan for up to 30 days.)

For example model [MA3.1.1] can be solved using Lindo to obtain results as follows:

Table 3.8.   Lindo Output on Model [MA3.1.1]

Objective value =		5600.00
Variable	**Value**	**Reduced cost**
PROD$_1$	160.00	
PROD$_2$		50.00
PROD$_3$		31.67
	Slack variables	**Shadow price**
GRIND	220.00	
POLISH		11.67
PACK	10.00	

This is interpreted as saying that the maximum weekly sales revenue is $5600 and the firm should only produce 160 units of product 1 and none of the other two products. The 'reduced costs' indicate that Products 2 and 3 will just become worth making if their contributions to sales revenue rise by $50 and $31.67 per unit respectively. Finally the 'slack values' indicate that there is spare capacity of 220 and 10 hours per week in Grinding and Packing respectively while all polishing time available is used up. The 'shadow price' of polishing suggests that at the margin, one hour of this process is worth $11.67 to the firm. For a more extensive explanation of these interpretations of solutions to linear programming models see Winston (1994) and Adnerson et al. (1999).

The Dual to a Linear Programming Model

Every linear programming model has a *dual* model associated with it. The optimal solution of the dual model can be deduced from that of the original model, known as the *primal* model. The dual model offers certain insights into the solution of the primal model most notably in terms of the implicit values of the resources giving rise to the primal's constraints. We introduce here the concept of duality in linear programming as it has relevance to DEA. As it will be seen in Chapter 4, linear programming duality makes it possible to switch from a production to a value-based context of efficiency assessment and vice versa.

The dual can be derived from the primal model after first putting the primal model in a *standard* form. There are alternative standard forms (e.g. maximisation, minimisation) which can be used. We cover here the case where the primal is put in a maximisation standard form.

Any linear programming model can be written in the following form:

Table 3.9.   Primal Model in Standard Maximisation Form

		Dual Variable	[MA3.1.2]
Max	$c_1x_1 + c_2x_2 + c_3x_3 + ... + c_nx_n$		
Subject to:			
	$a_{11}x_1 + a_{12}x_2 + ... + a_{1n}x_n \leq b_1$	y_1	
	$a_{21}x_1 + a_{22}x_2 + ... + a_{2n}\ x_n \leq b_2$	y_2	
	....		
	$a_{m1}x_1 + a_{m2}x_2 + ... + a_{mn}x_n \leq b_m$	y_m	

We have associated a *dual variable* with each one of the constraints to the primal model. The 'decision variables' of the dual model will be the dual variables. The RHS values of the primal model are used to construct the objective function of the dual model. The coefficients of each decision variable of the primal model are used to construct a constraint of the dual model. Finally the objective function coefficients of the primal model give rise to the RHS values of the dual model. The precise construction of the dual model is demonstrated below. The dual to [MA3.1.2] is [MA3.1.3]:

Table 3.10.   Dual to Model [MA3.1.2]

Min $b_1y_1 + b_2y_2 + b_3y_3 + ... + b_my_m$ Primal variable   [MA3.1.3]
Subject to:

$$a_{11}y_1 + a_{21}y_2 + ... + a_{m1}y_m \geq c_1 \qquad x_1$$
$$a_{12}y_1 + a_{22}y_2 + ... + a_{m2}y_m \geq c_2 \qquad x_2$$
....
$$a_{1n}y_1 + a_{2n}y_2 + ... + a_{mn}y_m \geq c_m \qquad x_n$$

The optimal solution of the dual model can be derived directly from that of the primal and vice versa. Coverage of these relationships is beyond the scope of this book but suffice it say that

the optimal objective function values of the primal and of the dual model are equal.

The Link Between DEA and Linear Programming

Linear programming was developed as an analytical method to aid decision making long before DEA was. It is a general purpose approach to reaching optimal decisions where a single objective is being pursued (e.g. the maximisation of profit or the minimisation of cost) and where a multitude of resource and policy constraints impact the decision to be made. Where linear programming and DEA come together is in that linear programming can be used as a tool to operationalise the concepts of efficiency measurement such as the mathematical definition of the PPS, the identification of Pareto-efficient DMUs and the computation of measures of the efficiency of DMUs.

As we can see in Chapter 3 the axioms underlying the transformation of inputs to outputs (see Appendix 3.2) make it possible to construct a set of constraints to a linear programming model so as to define feasible in principle input-output correspondences whether observed at DMUs or not. The objective function to the linear programming model can then be used in a variety of ways. In Chapter 3 it is used to give measures of the technical input or output efficiency of a DMU. As will be seen in Chapter 9, the objective function can be used to yield other measures of the efficiency of a DMU, target input-output levels and so on.

APPENDIX 3.2: POSTULATES FOR CONSTRUCTING THE PRODUCTION POSSIBILITY SET IN DEA UNDER CONSTANT RETURNS TO SCALE

Let DMU j use input levels $x_j \in \mathfrak{R}_+^m$, to produce output levels $y_j \in \mathfrak{R}_+^s$. The production possibility set PPS within which DMU j operates is denoted P and is defined as follows:

$P = \{ (x, y) \mid x \in \mathfrak{R}_+^m$ can produce $y \in \mathfrak{R}_+^s \}$.

The following assumptions are made to construct P, (see also Banker et al. (1984) p. 1081.)

i. Strong free disposability of input
If $(x', y') \in P$ and $x \geq x'$, then $(x, y') \in P$ where $x \geq x'$ means that at least one element of x is greater than the corresponding element of x'.
We referred 'informally' to this as a phenomenon of inefficient production.

ii. Strong free disposability of output
If $(x', y') \in P$ and $y \leq y'$, then $(x', y) \in P$.
We referred 'informally' to this as a phenomenon of inefficient production.

For a fuller discussion of strong disposability of input and output and the introduction of the concept 'weak' disposability of input and output see Cooper et al. (2000), section 3.14.

iii. No output can be produced without some input
$(x', 0) \in P$; but if $y' \geq 0$ then $(0, y') \notin P$.

iv. Constant returns to scale
If $((x', y')) \in P$ then for each positive real value $\lambda > 0$ we have $(\lambda x', \lambda y') \in P$.

v. Minimum extrapolation
All observed DMUs $\{(x_j, y_j), j = 1, 2,...N\} \in P$ and P is the smallest closed and bounded set satisfying i-iv.

A PPS P which satisfies the above postulates can be constructed from the observed DMUs (x_j, y_j), $(j = 1 ... N)$ as follows.
$\mathbf{P = \{(x, y) \mid x \geq \sum_j \lambda_j x_j; y \leq \sum_j \lambda_j y_j; \lambda_j \geq 0 \ \forall j\}.}$

Chapter 4

DATA ENVELOPMENT ANALYSIS UNDER CONSTANT RETURNS TO SCALE: GENERAL MODELS

4.1. INTRODUCTION

This chapter presents the basic DEA models for assessing technical input and output efficiency in the most general multi-input multi-output contexts. The models are referred to as *basic* to distinguish them from further DEA models developed in later chapters which can be seen as extensions to the models presented in this chapter.

The chapter begins with a formal generalisation to the multi-input multi-output case of the DEA models presented in Chapter 3. It then proceeds to present an alternative version of the general purpose DEA models which focuses on a value as distinct from a production framework for assessing DEA efficiency. The value-based definition of DEA efficiency offers additional insights into the performance of a DMU and may indeed be the only meaningful way to compare DMUs in certain real life situations.

4.2. A GENERAL PURPOSE LINEAR PROGRAMMING MODEL FOR ASSESSING TECHNICAL INPUT EFFICIENCY

Let us assume that we have N DMUs ($j = 1...N$) using m inputs to secure s outputs. Let us denote x_{ij} and y_{rj} the level of the ith input and rth output respectively observed at DMU j. The technical input efficiency of DMU j_0 is the optimal value of k_0 in Model [M4.1], first developed in Charnes et al. (1978). The model represents a generalisation to the multi-input multi-output case of the model in [M3.4] of Chapter 3. Let us use the superscript * to denote the optimal value of a variable in [M4.1]. Any feasible set of λ values in [M4.1] identifies a point within the PPS which can be constructed from DMUs ($j = 1...N$), using the postulates detailed in Appendix 3.2. Given that

$\varepsilon \ll 1$, [M4.1] gives pre-emptive priority to the minimisation of k_0. Thus the model identifies a point within the PPS which uses the lowest proportion k_0^* of the input levels of DMU j_0 while offering output levels which are at least as high as those of DMU j_0.

Thus k_0^* is by definition the technical input efficiency of DMU j_0.

Min $\qquad k_0 - \varepsilon[\sum_{i=1}^{m} S_i^- + \sum_{r=1}^{s} S_r^+]$ $\qquad\qquad$ [M4.1]

Subject to:

$$\sum_{j=1}^{N} \lambda_j \, x_{ij} = k_0 \, x_{ij_0} - S_i^- \qquad i = 1...m$$

$$\sum_{j=1}^{N} \lambda_j \, y_{rj} = S_r^+ + y_{rj_0} \qquad r = 1...s$$

$$\lambda_j \geq 0, j = 1...N, S_i^-, S_r^+ \geq 0 \, \forall \, i \, and \, r, \, k_0 \text{ free.}$$

ε is a non-Archimedean infinitesimal.

Once k_0 has been minimised the model seeks the maximum sum of the slack values S_i^- and S_r^+. If any one of these values is positive at the optimal solution to the model it means that the corresponding input or output of DMU j_0 can improve further, after its input levels have been contracted to the proportion k_0^*. Note that one feasible value of k_0 in [M4.1] is 1 as we can always set $\lambda_{j_0} = 1$, $\lambda_j = 0 \, \forall \, j \neq j_0$. Thus at the optimal solution to [M4.1] we will have $k_0^* \leq 1$.

If $k_0^* = 1$ and $S_r^{+*} = 0, r = 1....s$, $S_i^{-*} = 0, i = 1...m$ then DMU j_0 is Pareto-efficient because the model has been unable to identify some feasible production point which can improve on some input or output level of DMU j_0 without detriment to some other input or output level. In summary, the solution to [M4.1] yields the following information:

– DMU j_0 is Pareto-efficient if and only if $k_0^* = 1$ and $S_r^{+*} = 0, r = 1....s$, $S_i^{-*} = 0, i = 1...m$;

– DMU j_0 has technical input efficiency equal to k_0^*.

In practice, because of the difficulty of using a specific value for ε in [M4.1] the model is solved in two phases. During the first phase k_0 is minimized, ignoring the slack values. This yields the minimum value k_0^* of k_0. Then setting within [M4.1] $k_0 = k_0^*$ the model is solved to maximise $\sum_{i=1}^{m} S_i^- + \sum_{r=1}^{s} S_r^+$. This second phase yields an optimal set of slack values with assurance that k_0^* will not be reduced in any exchange for an increase in the slacks however large the latter might be, and however small is the reduction in k_0^*. (The optimal set of slack values need not be unique. Further, unless slack values are normalised in some way, the optimal slack values will depend on the scale in which input and output variables are measured.) This two-phase solution process is implemented in software normally including the *Warwick DEA software* accompanying this book. See also Ali in Charnes et al. (1994) on the issue of the non-Archimedean ε in DEA.) The following example illustrates the use of model [M4.1].

Example 4.1: A mail order company operates five regional distribution centres taking goods to customers. Table 4.1 shows data on the operations of the centres during the latest complete quarter.

Table 4.1.   Input-Output Data of Distribution Centres

Centre	Labour Hours (000)	Capital Employed ($m)	Delivery distance covered (000 km)	Packages delivered (000)
1	4.1	2.3	4.3	90
2	3.8	2.4	3.9	102
3	4.4	2.0	4.1	96
4	3.2	1.8	5.2	110
5	3.4	3.4	4.2	120

Set up a linear programming model to assess the technical input efficiency of Centre 1.

Solution: The inputs in this case are labour and capital. The outputs are distance covered and packages delivered, reflecting the volume of work done. The instance of the generic model [M4.1] corresponding to Centre 1 is model [M4.2] where the optimal value of k is the technical input efficiency of Centre 1, reflecting the maximum feasible radial contraction to its labour and capital levels of 4.1 and 2.3 respectively, without reducing its levels of distance and packages, standing at 4.3 and 90, respectively. If the model yields a positive optimal value for at least one of the slack variables SL, SC, SD and SP then that would be an improvement to the corresponding input or output level(s) feasible after the Centre's labour and capital levels have been reduced to the proportion represented by the optimal value of k. If at the optimal solution to model [M4.2] we have k = 1 and all slack values are zero then Centre 1 is Pareto-efficient.

Min $\quad$ k- ε (SL + SC + SD + SP) $\hfill$ [M4.2]

Subject to:

Labour: $\quad$ $4.1k - SL = 4.1\lambda_1 + 3.8\lambda_2 + 4.4\lambda_3 + 3.2\lambda_4 + 3.4\lambda_5$

Capital: $\quad$ $2.3k - SC = 2.3\lambda_1 + 2.4\lambda_2 + 2.0\lambda_3 + 1.8\lambda_4 + 3.4\lambda_5$

Distance: $\quad$ $4.3 + SD = 4.3\lambda_1 + 3.9\lambda_2 + 4.1\lambda_3 + 5.2\lambda_4 + 4.2\lambda_5$

Packages: $\quad$ $90 + SP = 90\lambda_1 + 102\lambda_2 + 96\lambda_3 + 110\lambda_4 + 120\lambda_5$

$\qquad$ $\lambda_1, \lambda_2, \lambda_3, \lambda_4, \lambda_5$, SL, SC, SD, SP ≥ 0, k free,

$\qquad$ ε is a non-Archimedean infinitesimal.

4.3. A GENERAL PURPOSE LINEAR PROGRAMMING MODEL FOR ASSESSING TECHNICAL OUTPUT EFFICIENCY

This model is readily constructed as a modification to the generic model [M4.1] for assessing technical input efficiency. Let us assume that we have the same N DMUs ($j = 1...N$) using m inputs to secure s outputs and let x_{ij} and y_{rj} be as defined in model [M4.1]. The technical output efficiency of DMU j_0 equals $1/h_{j_0}^*$ where $h_{j_0}^*$ is the optimal value of h_{j_0} in

Max $\qquad$ $h_{j_0} + \varepsilon [\sum_{i=1}^{m} I_i + \sum_{r=1}^{s} O_r]$ $\hfill$ [M4.3]

Subject to:

$$\sum_{j=1}^{N} \alpha_j x_{ij} = x_{ij_0} - I_i \qquad i = 1...m$$

$$\sum_{j=1}^{N} \alpha_j y_{rj} = O_r + h_{j_0} y_{rj_0} \qquad r = 1...s$$

$\alpha_j \geq 0, j = 1...N, I_i, O_r \geq 0 \forall i \, and \, r$, h_{j_0} free.

ε is a non-Archimedean infinitesimal.

We will use the superscript * to denote the optimal value of a variable in [M4.3]. The model gives pre-emptive priority to the maximisation of h_{j_0}. Thus it identifies a point within the PPS which offers output levels reflecting the maximum feasible radial expansion of the output levels of DMU j_0, without raising any one of its input levels. Thus,

by definition, $1/h_{j_0}^*$ is the technical output efficiency of DMU j_0.

The slack variables I_i and O_r in model [M4.3] are interpreted in a similar manner to the slack variables in [M4.1]. They represent any additional output augmentations and/or input reductions feasible from the PPS point corresponding to the maximum radial expansion of the output levels of DMU j_0 by the factor $h_{j_0}^*$. Note that one feasible value of h_{j_0} in [M4.3] is 1 as we can always set $\alpha_{j_0} = 1$, $\alpha_j = 0 \; \forall \; j \neq j_0$. Thus at the optimal solution to [M4.3] we will have $h_{j_0}^* \geq 1$.

If $h_{j_0}^* = 1$ and $O_r^* = 0, r = 1....s, \; I_i^* = 0, \; i = 1...m$ then DMU j_0 is Pareto-efficient. The model has been unable to identify a feasible production point which can improve on any one of the input or output levels of DMU j_0 without detriment to some other input or output level. In practice, Model [M4.3] is solved by a two-phase process as was the case with model [M4.1] earlier. During the first phase h_{j_0} is maximised, ignoring the slack values. This yields the maximum value $h_{j_0}^*$ of h_{j_0}. Then setting within [M4.3]

$$h_{j_0} = h_{j_0}^* \quad \text{the model is solved to maximise} \quad \sum_{i=1}^{m} I_i + \sum_{r=1}^{s} O_r .$$

In summary, the solution to [M4.3] yields the following information:
- DMU j_0 is Pareto-efficient if $h_{j_0}^* = 1$ and

$$O_r^* = 0, r = 1....s, \; I_i^* = 0, \; i = 1...m;$$

- DMU j_0 has technical output efficiency equal to $1 / h_{j_0}^*$.

It can be shown, see Cooper et al. (2000) section 3.8 that:
The technical input efficiency measure k_0^* yielded by model [M4.1] and the technical output efficiency measure $1 / h_{j_0}^*$ yielded by [M4.3] in respect of DMU j_0 are equal.

As models [M4.1] and [M4.3] yield equal measures of efficiency under constant returns to scale, either one is sufficient for estimating the technical input or output efficiency of a DMU. However, in cases where inputs are controllable while outputs are exogenous it is more sensible to use model [M4.1] which estimates technical input efficiency. In contrast, when outputs are controllable and inputs are exogenous it makes sense to use model [M4.3].

The next chapter will elaborate on the use and interpretation of DEA models in practice, including the identification of targets and role-model efficient DMUs. We conclude this section with an example illustrating the use of model [M4.3].

Example 4.2: Let us reconsider the mail order company of Example 4.1. Set up a linear programming model to assess the technical output efficiency of Centre 1.

Solution: We have already identified labour and capital as the inputs and distance covered and packages delivered as the outputs in the case of the centres being assessed. The instance of the generic model [M4.3] corresponding to Centre 1 is as follows:

Max $\qquad$ $H + \varepsilon (IL + IC + OD + OP)$ $\qquad$ [M4.4]

Subject to:

Labour: $\qquad$ $4.1 - IL = 4.1\alpha_1 + 3.8\alpha_2 + 4.4\alpha_3 + 3.2\alpha_4 + 3.4\alpha_5$

Capital: $\qquad$ $2.3 - IC = 2.3\alpha_1 + 2.4\alpha_2 + 2.0\alpha_3 + 1.8\alpha_4 + 3.4\alpha_5$

Distance: $\qquad$ $4.3 H + OD = 4.3\alpha_1 + 3.9\alpha_2 + 4.1\alpha_3 + 5.2\alpha_4 + 4.2\alpha_5$

Packages: $\qquad$ $90 H + OP = 90\alpha_1 + 102\alpha_2 + 96\alpha_3 + 110\alpha_4 + 120\alpha_5$

$\qquad$ $\alpha_1, \alpha_2, \alpha_3, \alpha_4, \alpha_5, IL, IC, OD, OP \geq 0,$

$\qquad$ H free, ε is a non-Archimedean infinitesimal.

The optimal value H* of H represents the maximum radial expansion possible of the levels of distance and packages of Centre 1, observed at 4.3 and 90 respectively. The expansion is without the need to raise its levels of labour and/or capital, observed at 4.1 and 2.3, respectively. This makes 1 / H* the technical output efficiency of Centre 1. If the model yields positive optimal values for any one of the slack variables IL, IC, OD and OP then they would reflect an improvement to the corresponding input or output feasible after Centre 1's distance and package levels have been expanded by the factor H*. If at the optimal solution to model [M4.4] we have H* = 1 and all slack values are zero then Centre 1 is Pareto-efficient.

4.4. VALUE-BASED DEA MODELS

We have introduced the concepts of efficiency in what could be said to be a *production* context. Measures of efficiency have been couched in terms of the potential for input conservation or output augmentation. We introduce in this section an equivalent context for the measurement of efficiency which is *value-based*. That is we measure efficiency by implicitly assigning values to inputs and outputs. The production-based DEA models we have covered so far, and most notably the generic models [M4.1] and [M4.3], will be referred to as *envelopment* models. The name reflects the fact that they measure DEA efficiency with reference to a PPS boundary which 'envelops' the input and output levels observed at DMUs as was illustrated graphically in Chapter 3. The DEA models we shall derive here with reference to the value-based context of efficiency measurement will be referred to as value-based DEA models. As we will see, the value-based DEA models are equivalent to the envelopment DEA models in the sense that each general purpose DEA model such as [M4.1] and [M4.3] has a *dual* model which measures DEA efficiency in a value context. (The concept of duality in linear programming was introduced in brief in Appendix 3.1 to Chapter 3.) The implicit values of inputs and outputs yielded by value-based DEA models contain important information on rates of substitution and transformation between the factors of production which enhances the usefulness of DEA analyses.

Let us consider the input-oriented envelopment model in [M4.1]. The equivalent value-based DEA model which is derived as the dual to [M4.1] is model [M4.5].

Max $\qquad \sum_{r=1}^{s} u_r y_{rj_0}$ $\hfill$ [M4.5]

Subject to:

$$\sum_{i=1}^{m} v_i x_{ij_0} = 1$$

$$\sum_{r=1}^{s} u_r y_{rj} - \sum_{i=1}^{m} v_i x_{ij} \leq 0 \qquad j = 1...j_0...N$$

$$u_r \geq \varepsilon \qquad\qquad\qquad r = 1...s$$

$$v_i \geq \varepsilon \qquad\qquad\qquad i = 1...m$$

where u_r and v_i are respectively the dual variables associated with the rth and ith constraint of [M4.1]. Notation in [M4.5] is otherwise as in [M4.1]. u_r and v_i are referred to as the <u>DEA-weight</u> on the rth output and ith input respectively. As [M4.5] was derived from the input-oriented envelopment model we shall refer to it as the 'input-oriented' value-based DEA model.

Let the superscript * denote the optimal value of the corresponding variable in [M4.5]. The technical input efficiency of DMU j_0 as yielded by the envelopment model [M4.1] was k_0^*. Let the optimal objective function value of [M4.5] be $h'_{j_0} = \sum_{r=1}^{s} u_r^* y_{rj_0}$. By virtue of duality (see Appendix 3.1) h'_{j_0} will also equal the optimal objective function value of [M4.1] which in view of the very small value of ε, virtually equals k_0^*.

Thus the measure of technical input efficiency yielded by [M4.5] is $h'_{j_0} = \sum_{r=1}^{s} u_r^* y_{rj_0}$ which is approximately, but not exactly, equal to the true technical input efficiency k_0^*.

The approximation is better the smaller the value of ε used in [M4.1] and [M4.5]. (ε cannot be set to zero, of course, for then we would not necessarily identify Pareto-efficient points by means of [M4.1] or [M4.5].)

DMU j_0 is Pareto-efficient if $\sum_{r=1}^{s} u_r^* y_{rj_0} = 1$ in [M4.5].

This can be deduced by noting that through duality of [M4.5] and [M4.1] we have $\sum_{r=1}^{s} u_r^* y_{rj_0} = k_0^* - \varepsilon[\sum_{i=1}^{m} S_i^{-*} + \sum_{r=1}^{s} S_r^{+*}] = 1$. Given the pre-emptive priority on the minimisation of k_0 in [M4.1] $k_0^* - \varepsilon[\sum_{i=1}^{m} S_i^{-*} + \sum_{r=1}^{s} S_r^{+*}] = 1$ implies that $k_0^* = 1$, $S_r^{+*} = 0$, $r = 1....s$, and $S_i^{-*} = 0$, $i = 1...m$. It was seen earlier that such a solution to [M4.1] means DMU j_0 is Pareto-efficient. (In practice, because of round-off errors it is safer to test for Pareto-efficiency using the two-phase process outlined in the context of solving the envelopment model [M4.1].)

The constraint $\sum_{i=1}^{m} v_i x_{ij_0} = 1$ in [M4.5] is known as a *normalisation constraint*. Its RHS value of 1 is arbitrary to the extent that in the objective function to the envelopment model [M4.1] the coefficient of k_0 can be set to any factor other than zero. It is advisable to set the coefficient to k_0 in [M4.1] and equivalently the RHS of the normalisation constraint in [M4.5] to some value larger than 1 to avoid round off errors is computing the DEA weights in that model. (Typically the value of 100 is used as a normalisation constant.)

The output-oriented envelopment model in [M4.3] also gives rise through duality to a value-based DEA model which is as follows:

Min $\qquad \sum_{i=1}^{m} \delta_i x_{ij_0}$ $\qquad\qquad\qquad\qquad$ [M4.6]

Subject to:

$$\sum_{r=1}^{s} \gamma_r y_{rj_0} = 1$$

$$\sum_{r=1}^{s} \gamma_r y_{rj} - \sum_{i=1}^{m} \delta_i x_{ij} \le 0 \qquad j = 1...j_0...N$$

$$\gamma_r \ge \varepsilon \qquad\qquad\qquad r = 1...s$$

$$\delta_i \ge \varepsilon \qquad\qquad\qquad i = 1...m.$$

We shall use the superscript * to denote the optimal value of the corresponding variable in [M4.6]. Let the optimal objective function value of [M4.6] be $z_{j_0}' = \sum_{r=1}^{s} \delta_i^* x_{ij_0}$. By virtue of duality, z_{j_0}' will also equal the optimal objective function value of [M4.3] which in view of the very small value of ε, virtually equals $h_{j_0}^*$. The technical output efficiency of DMU j_0 as yielded by the envelopment model [M4.3] was $1 / h_{j_0}^*$.

Thus the measure of technical output efficiency yielded by [M4.6] is $1 / z_{j_0}'$ which is approximately, but not exactly, equal to the true technical output efficiency $1 / h_{j_0}^*$.

The approximation is better the smaller the value of ε used in [M4.3] and [M4.6].

DMU j_0 is Pareto-efficient if $\sum_{r=1}^{s} \delta_i^* x_{ij_0} = 1$ in [M4.6].

This can be deduced by recourse to the duality between [M4.6] and [M4.3] in the manner duality between [M4.5] and [M4.1] was used above to establish that when $\sum_{r=1}^{s} u_r^* y_{rj_0} = 1$ in [M4.5] DMU j_0 is Pareto-efficient. (In practice, because of round-off errors, it is safer to test for Pareto-efficiency using the two-phase process outlined in the context of solving the envelopment model [M4.3].)

The constraint $\sum_{r=1}^{s} \gamma_r y_{rj_0} = 1$ in [M4.6] is the normalisation constraint mirroring that of model [M4.5]. In this case too the value of the RHS has arbitrarily been set to 1 just as the coefficient of h_{j_0} in [M4.3] was arbitrarily set to 1. It is advisable to set the coefficient of h_{j_0} in [M4.3] and equivalently the RHS of the normalisation constraint in [M4.6] to a larger value to avoid round off errors when computing the DEA weights in [M4.6].

4.5. INTERPRETATION OF VALUE-BASED DEA MODELS

The value-based DEA models add a new perspective on assessment of performance, defining as they do efficiency with reference to the (implicit) values of the inputs and outputs. Without loss of generality let us take the input-oriented value-based DEA model in [M4.5]. In order to see more clearly the way it arrives at the efficiency measure of DMU j_0 we shall convert the model to an equivalent *linear fractional* model. The model in [M4.5] yields the same objective function value h'_{j_0} as the linear fractional model in [M4.7].

$$h'_{j_0} = Max \frac{\sum_{r=1}^{s} U_r y_{rj_0}}{\sum_{i=1}^{m} V_i x_{ij_0}} \qquad [M4.7]$$

Subject to:

$$\frac{\sum_{r=1}^{s} U_r y_{rj}}{\sum_{i=1}^{m} V_i x_{ij}} \leq 1 \qquad j = 1...j_0...N$$

$$U_r \geq \varepsilon \qquad r = 1...s$$
$$V_i \geq \varepsilon \qquad i = 1...m.$$

U_r and V_i are the decision variables in [M4.7]. Notation in [M4.7] is otherwise as in [M4.5]. The equivalence between the linear fractional model [M4.7] and the linear programming model in [M4.5] is based on Charnes and Cooper (1962). It is noted that [M4.5] leads to the specific solution of [M4.7] in which the values of U_r and V_i are such that they lead to the denominator of the objective function of [M4.7] being equal to 1. An infinite set of other solutions is possible scaling up or down the values of U_r and V_i by the same positive constant. Such scaling of the decision variable values leaves the objective function to [M4.7] unaffected as the scaling constant of the numerator cancels that of the denominator of the objective function in [M4.7].

U_r can be seen as **the imputed value of the marginal unit** of output r. Similarly, V_i can be seen as **the imputed value of the marginal unit** of input i. The fractional form of the objective function of [M4.7] means that it is the *relative* rather than the absolute levels of these imputed values that are important. When we view U_r and V_i as imputed values of output r and input i respectively, we see that:

the efficiency measure h'_{j_0} of DMU j_0 yielded by [M4.7] is the ratio of the total imputed value of its output levels to the total imputed value of its input levels.

The use of ratios as measures of efficiency is not new to DEA. Ratios of output to input are well established in performance measurement as we saw in Chapter 1. In multi-input multi-output contexts practitioners typically use prior weights to reduce the multiple outputs to a single 'weighted output' and the multiple inputs to a single 'weighted input' to arrive at a single ratio measure of performance. What makes the DEA total weighted output to input ratio in model [M4.7] different are two key facts:
– No prior weights or values are imposed on inputs and outputs and;
– The input-output values are DMU-specific, chosen to maximise the efficiency rating of the respective DMU.
These two features of the imputed input-output values have important practical implications as will be seen later in this chapter and in Chapter 5.

As noted above only the relative rather than the absolute levels of the imputed input-output values yielded by value-based DEA models matter. These relative levels specify *the marginal rates of substitution* between inputs or between outputs and the *marginal rates of transformation* between inputs and outputs. For example it can be shown (see Charnes et al. (1978)) that if at the optimal solution to [M4.5] $v_i^* \neq \varepsilon$ and $v_j^* \neq \varepsilon$ then $\dfrac{dx_i}{dx_j} = -\dfrac{v_j^*}{v_i^*}$.

In practical terms this means that if [M4.5] yields non-ε weights of u_i^* and u_j^* for u_i and u_j respectively then $-u_i^* / u_j^*$ is the marginal rate of substitution between output i and j in that an additional unit of output i would be compensated for by a $-u_i^* / u_j^*$ drop in the level of output j.

Unfortunately, the above marginal rates of substitution and/or transformation cannot always be derived in DEA. Specifically, when one or more of the implicit input or output values yielded by a value-based DEA model is ε, rates of substitution and/or transformation involving the corresponding input or output cannot be defined. Moreover, the value-based DEA model may have multiple optimal solutions complicating still further the determination of *the* marginal rates of substitution and/or transformation between the factors of production. (For a fuller discussion of the interpretation of the implicit input-output values in value-based DEA models as marginal rates of substitution or transformation see Charnes et al. (1978).)

One of the practical uses to which DEA-based implicit input-output values can be put is in computing the '*virtual*' input and output levels of a DMU. For example let v_i^* and u_r^* be respectively the optimal values of v_i and u_r in model [M4.5]. Then

$x_{ij_0} v_i^*$ is the ith 'virtual' input and $y_{rj_0} u_r^*$, the rth 'virtual' output of DMU j_0.

Total virtual input is normalised to 1 in [M4.5] and so total virtual output is also 1 for Pareto-efficient DMUs. Thus *we can see the level of a virtual input or output as a normalised weight indicating the extent to which the efficiency rating of a Pareto-efficient DMU is underscored by that input or output variable*. We shall return to the practical usefulness of virtual input and output levels in Chapter 4.

Example 4.3: Set up the value-based DEA model to assess
a) the technical input efficiency and
b) the technical output efficiency
of Centre 1, Example 4.1.

Solution: a) We identified previously that in the context of the distribution centres of Example 4.1 the inputs are Labour and Capital while the outputs are Distance and Packages. Let u_D, u_P, v_L and v_C be respectively the values to be imputed to Distance, Packages, Labour and Capital when assessing Centre 1. Then the value-based DEA model for ascertaining the technical input efficiency of Centre 1 is as follows:

Max $\qquad Z_1 = 4.3u_D + 90u_P$ $\qquad\qquad$ [M4.8]
Subject to:
Normalisation: $4.1v_L + 2.3v_C = 1$
Centre 1: $\qquad 4.3u_D + 90u_P - 4.1v_L - 2.3v_C \leq 0$
Centre 2: $\qquad 3.9u_D + 102u_P - 3.8v_L - 2.4v_C \leq 0$
Centre 3: $\qquad 4.1u_D + 96u_P - 4.4v_L - 2v_C \leq 0$
Centre 4: $\qquad 5.2u_D + 110u_P - 3.2v_L - 1.8v_C \leq 0$
Centre 5: $\qquad 4.2u_D + 120u_P - 3.4v_L - 3.4v_C \leq 0$
$\qquad\qquad u_D, u_P, v_L, v_C \geq \varepsilon, 0 < \varepsilon, \varepsilon$ non-Archimedean infinitesimal.

Let Z_1^* be the optimal value of Z_1 in [M4.8]. The technical input efficiency of Centre 1 is Z_1^*. If $Z_1^* = 1$ then Centre 1 is Pareto-efficient.

b) The model for ascertaining the technical output efficiency of Centre 1 is as follows.

Min $\qquad H_1 = 4.1\delta_L + 2.3\delta_C$ $\qquad\qquad$ [M4.9]
Subject to:
Normalisation: $4.3\gamma_D + 90\gamma_P = 1$
Centre 1: $\qquad 4.3\gamma_D + 90\gamma_P - 4.1\delta_L - 2.3\delta_C \leq 0$
Centre 2: $\qquad 3.9\gamma_D + 102\gamma_P - 3.8\delta_L - 2.4\delta_C \leq 0$
Centre 3: $\qquad 4.1\gamma_D + 96\gamma_P - 4.4\delta_L - 2\delta_C \leq 0$
Centre 4: $\qquad 5.2\gamma_D + 110\gamma_P - 3.2\delta_L - 1.8\delta_C \leq 0$
Centre 5: $\qquad 4.2\gamma_D + 120\gamma_P - 3.4\delta_L - 3.4\delta_C \leq 0$
$\qquad\qquad \gamma_L, \gamma_P, \delta_L, \delta_2 \geq \varepsilon, 0 < \varepsilon, \varepsilon$ non-Archimedean infinitesimal.

Let H_1^* be the optimal value of H_1 in [M4.9]. The technical output efficiency of Centre 1 is $1 / H_1^*$. If $H_1^* = 1$ then Centre 1 is Pareto-efficient.

4.6. EFFICIENT PEERS AND TARGETS IN DEA

As indicated in previous chapters, apart from yielding a measure of the input or output efficiency of a DMU, in the case of inefficient DMUs DEA models also identify *target* input-output levels which would render them Pareto-efficient and *efficient peers* they could emulate to improve their performance. Targets and peers are of significant practical value and we discuss them in more detail here.

4.6.1 Targets

Consider the solution of model [M4.1] to assess the technical input efficiency of DMU j_0. We shall use the superscript * to denote the optimal values of variables in that model. The model identifies in respect of DMU j_0 a set of Pareto-efficient input-output levels (x_i^t, $i = 1...m$, y_r^t, $r = 1...s$) where,

$$x_i^t = \sum_{j=1}^{N} \lambda_j^* x_{ij} = k_0^* x_{ij_0} - S_i^{-*} \qquad i = 1...m$$

$$y_r^t = \sum_{j=1}^{N} \lambda_j^* y_{rj} = S_r^{+*} + y_{rj_0} \qquad r = 1...s \qquad (4.1).$$

The input-output levels (x_i^t, $i = 1...m$, y_r^t, $r = 1...s$) defined in (4.1) are the co-ordinates of the point on the efficient frontier used as a benchmark for evaluating DMU j_0. The proof that the input-output levels in (4.1) would render DMU j_0 Pareto-efficient can be found in Cooper et al. (2000) section 3.5. These levels are often referred to as a *projection* point of DMU j_0 on the efficient boundary or simply *targets* of DMU j_0.

When a DMU is Pareto-inefficient the input-output levels in (4.1) can be used as the basis for setting it targets so that it can improve its performance. There are generally infinite input-output level combinations which would render any given DMU Pareto-efficient. The specific combination in (4.1) corresponds to giving pre-emptive priority to the radial contraction of the input levels of DMU j_0. This means that the targets in (4.1) preserve in large measure the *mix* of inputs and outputs of DMU j_0, though if any slack values are positive the mix of the target input-output levels differs from that of DMU j_0. The preservation of the input-output mix in target setting has advantages. The mix typically reflects a combination of operating choices (e.g. the relative levels of labour and automation) and uncontrollable factors (e. g. the types of crime dealt with by a police force). Setting an inefficient DMU targets which preserve in large measure its own mix of inputs and outputs makes it easier for the DMU to accept them and attempt to attain them. This is because the targets will reflect in large measure its own operating priorities, history and environment. By the same token, however, by their very nature the targets in (4.1) will not be suitable in contexts where the aim is to encourage certain inefficient DMUs to alter operating priorities and/or move to new environments.

DEA targets have so far been discussed with reference to the input-minimising DEA model in [M4.1]. The output maximising DEA model in [M4.3] also yields targets and peers. We shall again use the superscript * to denote the optimal values of variables in that model. The model identifies in respect of DMU j_0 a set of Pareto-efficient input-output levels (x_i^t, $i =1...m$, y_r^t, $r=1...s$) where,

$$x_i^t = \sum_{j=1}^{N} \alpha_j^* \, x_{ij} = x_{ij_0} - I_i^* \qquad\qquad i = 1...m$$

$$y_r^t = \sum_{j=1}^{N} \alpha_j^* y_{rj} = O_r^* + h_{j_0}^* \, y_{rj_0} \qquad\qquad r = 1...s \qquad\qquad (4.2)$$

The key difference between the target levels in (4.1) and (4.2) is that those in (4.2) are arrived at by giving pre-emptive priority to the radial expansion of the output levels of DMU j_0. When radial expansion of outputs is complete the targets exploit further reductions in individual input and/or rises in individual output levels that are feasible at the radially expanded output levels. Such individual input savings and/or output augmentations are reflected in the slack values I_i^* and O_r^*.

Clearly the targets in (4.1) would be more suitable in cases where in the short term input levels are controllable while output levels are not. The targets in (4.2) are more suitable when the opposite is the case. Often in practice some of the inputs and some of the outputs are controllable while others are not. For such cases specially adapted DEA models have been developed by Banker and Morey (1986a) which are introduced in Chapter 9. For the time being, it is worth noting that if we use say the targets in (4.1) in a case where outputs are uncontrollable any output augmentations indicated in the targets can still be meaningful. Take for example water distribution where inputs may be various types of expenditure while the outputs may include the length of main used, an uncontrollable output (in the short term). Then any augmentation to the length of main within the targets in (4.1) can be interpreted as saying that even if length of main were to increase (e.g. through extensions to new housing estates) to the level indicated in the targets, the radially contracted input (expenditure level) should still be sufficient.

4.6.2 Efficient Peers

We can readily identify the efficient peers to a DMU whether we use an envelopment or a value-based DEA model. The two models bring to the fore different aspects of the significance of a peer.

Let us first consider one of the envelopment models, [M4.1], and let us use the superscript * to denote the optimal values of the corresponding variables in that model.

The *'efficient peers'* or *'efficient referents'* to DMU j_0 are those DMUs which correspond to positive λ^*'s.

We can see the practical significance of the efficient peers now if we look again at the targets in (4.1) which model [M4.1] yields for DMU j_0. It is clear that the target level for DMU j_0 on a given input is a linear combination of the levels of that input at its efficient peers. The same is true of the target level on each output for DMU j_0. Further, it is easy to deduce from (4.1) that

the efficiency rating k_0^* of DMU j_0 is the maximum of the ratios $\dfrac{x_i^t}{x_{ij_0}}$

$i = 1...m$. Thus the target input-output levels of DMU j_0 and its efficiency rating are exclusively dependent on the observed input-output levels of its efficient peers and on no other DMUs. (It is recalled that different priorities on input-output improvements (and indeed other DEA models to be covered in Chapter 9) will lead generally to different efficient peers.)

Consider now using one of the value-based DEA models, say [M4.5], to assess DMU j_0.

Its efficient peers are the DMUs corresponding to constraints which are binding at its optimal solution.

These are of course the same efficient peers as would be identified had model [M4.1] been used to assess DMU j_0 instead. [This can be shown to be true by recourse to the duality between models [M4.1] and [M4.5], covered in most textbooks on linear programming.]

It was noted earlier in this chapter that one of the features of the DEA weights is that they represent imputed input-output values which are DMU-specific, chosen to maximise the efficiency rating of DMU j_0. Thus the optimal DEA-weights of model [M4.5] favour the mix of input-output levels of DMU j_0 and since they render its peers efficient, the latter must have similar mixes of input-output levels to DMU j_0. It is recalled that the term mix of input-output levels refers to the ratios the input-output levels of a DMU are to each other. Thus the efficient peers have a 'similar' (if not identical) mix of input-output levels to that of DMU j_0 but at more efficient absolute levels. That is its efficient peers generally will have lower input relative to output levels than DMU j_0.

The features of efficient peers we have identified make them very useful in practice as role models DMU j_0 can emulate so that it may improve its performance. They are efficient and given that they have similar mix of input-output levels to DMU j_0 they are likely to operate in similar environments and/or to favour similar operating practices to DMU j_0.

4.7. INPUT AND OUTPUT ALLOCATIVE EFFICIENCIES

We saw in Chapter 1 that when input prices are available it would be valuable to compute the *input allocative efficiency* or price efficiency of DMUs. This measures the component of any cost inefficiency of a DMU which is attributable to its use of an uneconomic mix of inputs, given the input prices it faces. Similarly, when output rather than input prices are available we may compute *output allocative efficiency* which measures the component of any revenue shortfall of a DMU which is attributable to its procuring an inappropriate mix of outputs in light of the prices of the outputs. These concepts of cost minimising and revenue maximising efficiencies are described here along with models for computing the related allocative efficiency measures. (See also Chapter 8 in Cooper et al. (2000) and Chapter 3 in Fare et al. 1994b on models for assessing allocative, profit and revenue efficiencies of DMUs.)

4.7.1 Cost Minimising Efficiencies

Let us have DMUs ($j = 1...N$), the jth one using inputs x_{ij} ($i = 1...m$), to secure outputs y_{rj} ($r = 1...s$). Let the jth DMU face input prices w_{ij} ($i = 1...m$). We need to compute C_{j_0}, which is the least cost at which DMU j_0 could have produced its outputs y_{rj_0} ($r = 1...s$). C_{j_0} is derived by solving the following model:

$$C_{j_0} = \underset{x_i}{Min} \sum_{i=1}^{m} w_{ij_0} x_i \qquad\qquad \text{[M4.10]}$$

Subject to:

$$\sum_{j=1}^{N} \lambda_j x_{ij} \le x_i \qquad\qquad i = 1...m$$

$$\sum_{j=1}^{N} \lambda_j y_{rj} \ge y_{rj_0} \qquad\qquad r = 1...s$$

$$\lambda_j \ge 0, j=1...N \ge 0, \ x_i \ge 0, \ \forall i.$$

The optimal solution to model [M4.10] identifies the input levels x_i^*, ($i = 1..m$), DMU j_0 could have used in light of the input prices it faces, in order to secure its output levels at least cost. Then

the *input overall efficiency* of DMU j_0 is $IOE_{j_0} = \dfrac{C_{j_0}}{OC_{j_0}}$ (4.3),

where $OC_{j_0} = \sum_{i=1}^{m} w_{ij_0} x_{ij_0}$ is the cost at which DMU j_0 delivers its output.

The input overall efficiency IOE_{j_0} of DMU j_0 measures the extent to which its observed cost OC_{j_0} can be reduced given its input prices and output levels. When $IOE_{j_0} < 1$ the cost at which DMU j_0 delivers its outputs is in excess of the minimum required. This may be due to using excess amounts of input or/and using the wrong input mix in light of input prices. The first form of inefficiency is captured by the technical input efficiency and the second is captured residually by the input oriented measure of *allocative* (or *price*) *efficiency*, defined as follows.

Let k_0^* be the technical input efficiency of DMU j_0 obtained by solving model [M4.1] and let IOE_{j_0} be its input overall efficiency defined in (4.3). Then

the *input allocative efficiency* of DMU j_0 is $IAE_{j_0} = \dfrac{IOE_{j_0}}{k_0^*}$ (4.4)

When $\mathrm{IAE}_{j_0} = 1$ DMU j_0 has equal input technical and overall efficiencies. This means it is using the most economic mix of inputs in light of the prices at which it can secure them. When $\mathrm{IAE}_{j_0} < 1$ DMU j_0 is not using the most economic mix of inputs. Even if its technical input efficiency is 1, and so the DMU is using as low input levels as possible for its output levels, the mix of the inputs is such that the total cost at which it secures its outputs is higher than it need be. Model [M4.10] above identifies the input levels which would render DMU j_0 both technically and allocatively efficient in the input cost minimising sense. (For alternative (*additive*) models for estimating technical and allocative efficiencies and marginal rates of substitution see Cooper et al. (2000b))

4.7.2 Revenue Maximising Efficiencies

The approach to assessing input allocative efficiency can be readily adapted to assessing overall and allocative efficiency when output rather than input prices are available. This time the measures would relate to revenue maximising behaviour.

Consider the N DMUs referred to in connection with model [M4.10] and let the jth DMU face output prices p_{rj} ($r = 1...s$). We need to compute R_{j_0}, which is the maximum revenue DMU j_0 could have secured for its inputs x_{ij_0} ($r = 1...m$). R_{j_0} is derived by solving the following model:

$$R_{j_0} = \underset{y_r}{\mathrm{Max}} \sum_{r=1}^{s} p_{rj_0} y_r \qquad [M4.11]$$

Subject to:

$$\sum_{j=1}^{N} \lambda_j x_{ij} \le x_{ij_0} \qquad i = 1...m$$

$$\sum_{j=1}^{N} \lambda_j y_{rj} \ge y_r \qquad r = 1...s$$

$$\lambda_j \ge 0, j=1...N \ge 0, \ x_i \ge 0, \forall i.$$

The optimal solution to model [M4.11] identifies the output levels y_r^* ($r = 1...s$), DMU j_0 could have secured in order to maximise its revenue, given the output prices it faces. Then

the *output overall efficiency* of DMU j_0 is $\mathrm{OOE}_{j_0} = \dfrac{\mathrm{OR}_{j_0}}{R_{j_0}}$ (4.5), where $OR_{j_0} = \sum_{i=1}^{s} p_{rj_0} y_{rj_0}$ is the observed revenue of DMU j_0.

The overall revenue efficiency OOE_{j_0} of DMU j_0 measures the extent to which its observed revenue OR_{j_0} can be raised given its input levels. When $OOE_{j_0} < 1$ the revenue of DMU j_0 is short of what it could be. This may be due to producing lower than feasible output levels or/and producing the wrong output mix in light of the output prices it faces. The first form of inefficiency is captured by the technical output efficiency and the second is captured residually by the *output allocative efficiency* measure defined as follows.

Let $1 / h_{j_0}^{*}$ be the technical output efficiency of DMU j_0 obtained by solving model [M4.3] and let OOE_{j_0} be its output overall efficiency defined in (4.5). Then

the *output allocative efficiency* of DMU j_0 is $OAE_{j_0} = OOE_{j_0} h_{j_0}^{*}$ (4.6)

When $OAE_{j_0} = 1$ DMU j_0 has equal output technical and overall efficiencies. This means it is securing the best revenue yielding mix of outputs in light of the output prices it faces. When $OAE_{j_0} < 1$ DMU j_0 is not producing the best mix of output levels for the output prices it faces. Even if its technical output efficiency is 1, meaning the DMU is securing maximum output levels for its input levels, the mix of the outputs is such that the total revenue they generate is less than it could be. Model [M4.11] above identifies the ouput levels which would render DMU j_0 both technically and allocatively efficient in the revenue maximising sense.

4.8. QUESTIONS

1.   Reconsider the branches of the bank in Question 1 of section 8 in Chapter 3. Set up the value-based DEA model to ascertain the technical input efficiency of Branch 1. Solve the model and ascertain the imputed values of the inputs and outputs which give Branch 1 its maximum efficiency rating. Are these imputed values unique?

2.   Solve the value-based DEA model to assess the technical output efficiency of hospital E in Table 3.3 of Chapter 3. Compare the result obtained with that derived by solving the envelopment model [M3.5] in Chapter 3, explain any discrepancies in the values of the technical output efficiency obtained. What are the values of the DEA weights which give hospital E its maximum efficiency rating? Are they unique?

3.   Consider Centre 2 in Table 4.1.
   i.   Set up the DEA envelopment model to assess its technical input efficiency.
   ii.   Set up the DEA envelopment model to assess its technical output efficiency.
   iii. Set up the DEA value-based model to assess its technical input efficiency.
   iv.   Set up the DEA value-based model to assess its technical output efficiency.
   v   Solve the model set up in (iii) and interpret the DEA weights obtained.
   vi.   Contrast the efficiency estimates obtained for Centre 2 by the four alternative DEA models set in (i)-(iv).

4.   Reconsider the local tax offices of question 4 in Section 8 of Chapter 3 where management wishes to add *capital charge* as a second input to that of operating cost to reflect assets such as computers and physical storage and retrieval equipment in use at each office. The data for the most recent financial year is as follows:

Table 4.2.   Data on Tax Offices

Office	Running Cost	Capital Charge	No of accounts	No of adjunt applns	No of Sum monses	NPV of taxes
	($00,000)	($0,000)	(0,000)	(000)	(000)	($10m)
1	9.13	8.12	7.52	34.11	21.96	3.84
2	13.60	11.40	8.30	23.27	46.00	8.63
3	5.65	6.30	1.78	0.16	1.31	39.01
4	12.71	10.90	6.32	13.63	8.53	5.16
5	11.19	12.40	6.58	10.90	3.52	3.46

In respect of office 3:

i.   Set up the DEA envelopment model to assess its technical input efficiency.

ii.   Set up the DEA envelopment model to assess its technical output efficiency.

iii. Set up the DEA value-based model to assess its technical input efficiency.

iv.   Set up the DEA value-based model to assess its technical output efficiency.

v   Solve the model set up in (iii) and interpret the DEA weights obtained.

vi.   Contrast the efficiency estimates obtained for office 3 by the four alternative DEA models set in (i)-(iv).

5.   An insurance company employs six salespersons to serve various regions of the country. The products sold by the salespersons are life policies, pension plans and savings plans. The company has the following data on the activities of the salespersons in the latest quarter financial year.

Table 4.3.   Hours of Work and Financial Products Sold

Sales person	Hours worked (00)	Life Policies	Pension Plans	Savings Plans
1	4.2	50	30	45
2	4	35	40	50
3	4.7	80	35	40
4	4.3	20	50	30
5	3.8	53	25	35
6	5	55	33	74

i.   Set up and solve the value-based and the envelopment model to assess the technical input efficiency of salesperson 3.

ii.   In solving the envelopment model to assess the technical output efficiency of salesperson 1 it is found that the optimum value of the objective function is 1.09 and all slack variables are zero.
a)   Is salesperson 1 relatively efficient? Why?
b)   How do you interpret the optimum objective function value in terms of the sales transactions salesperson 1 secured during the latest quarter financial year?

iii.   Discuss whether the factors on which data was given above are adequate to assess the salespersons on their relative efficiency in selling the company's products.

6.   Reconsider the salespersons in Question 5 and let us assume that each salesperson was assisted by telephone staff of the insurance company making preparatory telephone calls to set up sales meetings between the salespersons and potential clients. Assume further that the cost per hour for salespersons and telephone staff differs in the six regions covered by the salespersons. The data is as follows.

Table 4.4.   Revised Data for the Salespersons of Question 5

Sales-person	Hours worked (00)		Hourly cost ($)		Sales of financial products		
	Tele-staff	Sales-person	Teleph staff	Sales-person	Life Pol'ies	Pen'n Plans	Sav's Plans
1	0.8	4.2	5	12	50	30	45
2	1.1	4	4.5	15	35	40	50
3	1.25	4.7	3.8	16	80	35	40
4	0.7	4.3	3.2	18	20	50	30
5	0.76	3.8	4.2	14	53	25	35
6	1.4	5	1.8	13	55	33	74

Assess the input allocative (price) efficiency of salesperson 5. Which salesperson(s) has (have) the highest input allocative efficiency?

7.   Reconsider question 6 and assume the cost per hour of sales persons and telephone staff is not available. Instead, we have the long term revenue the insurance company expects to gain from the financial products sold. Moreover, assume that these revenues differ by region because of the different rates of customer retention and investment levels in the different regions covered by the salespersons. The data is as follows.

Table 4.5.   Data on Salespersons when Unit Revenues are Available

Sales-per'n	Hours worked (00)		Transactions completed			Revenue per transaction ($000)		
	Tel. staff	Sales-per'n	Life Pol'es	Pens. Plans	Sav's Plans	Life Pol'es	Pens. Plans	Sav's Plans
1	0.8	4.2	50	30	45	4	3	1
2	1.1	4	35	40	50	2.2	2.8	1.3
3	1.25	4.7	80	35	40	3.6	3.2	1.1
4	0.7	4.3	20	50	30	2.3	2.9	1.2
5	0.76	3.8	53	25	35	2.5	3.1	1.15
6	1.4	5	55	33	74	2.3	2.8	2

Assess the overall and allocative efficiency of salesperson 1 from the revenue maximising perspective.

Chapter 5

USING DATA ENVELOPMENT ANALYSIS IN PRACTICE

5.1 INTRODUCTION

In this chapter we focus on the practicalities of carrying out an assessment of comparative performance by means of DEA. In this context we cover the principles governing the choice of input-output variables, the type of information on performance most often retrieved in the course of a DEA assessment, the use of DEA software for carrying out assessments of performance and some additional advice on carrying out DEA assessments.

5.2 CHOOSING INPUTS AND OUTPUTS IN A DEA ASSESSMENT

As already noted in earlier chapters, the identification of the input-output variables to be used in an assessment of comparative performance is the first and arguably the most important stage in carrying out the assessment. The results obtained depend crucially on the choice made. The input-output variables are naturally unique to the type of efficiency being assessed, (e.g. operating efficiency in delivering goods versus effectiveness in adding value in some process such as education), the context in which the units to be assessed operate, and the factors which are exogenous for the echelon of management being assessed. This makes it difficult to be specific on what the input-output variables should be even when the context of the assessment is known. We focus, instead therefore, on the principles governing the choice of input-output variables and leave it to the user to apply them to the specific context in which they are proposing to carry out a DEA assessment.

As we saw in Chapter 4, at the very heart of an assessment of comparative efficiency by DEA lies the construction of the Production Possibility Set (PPS) containing all input-output level "correspondences" which are capable of being observed. Correspondence of inputs and outputs in this context is based on a relationship of *exclusivity* and *exhaustiveness* between the two sets of variables.

The relationship of exclusivity and exhaustiveness between inputs and outputs in a DEA assessment means that subject to the exogeneity of factors involved, the inputs and they alone must influence the output levels, and only of the outputs being used in the assessment[1].

The identification of the variables which are exogenous is important. For example if in assessing a sales outlet we wish to ascertain its effectiveness in attracting custom and so use sales revenue as an output variable, it would not be wise to use as an input the staff complement or expenditure on staff if staff levels can be controlled and therefore are not exogenous to the outlet management. Thus though staff numbers do impact sales revenue, an output variable being used, we do not use staff as an input for we would risk showing an outlet efficient when it generates low sales revenue despite the potential in its locality, simply by virtue of its choosing to employ low staff numbers. Thus exclusivity and exhaustiveness of input-output variables must guide the choice of the input-output variables subject to the exogeneity of any variables being proposed.

The identification of the inputs and the outputs and the establishment of exclusivity and exhaustiveness between them is in practice very difficult. A starting point is the establishment of the type of efficiency to be assessed. This could differ even when we are concerned with the same set of outlets. For example a sales outlet may be assessed on its operating efficiency where operating expenditure is to be judged against activity volumes, or it may be assessed on its ability to attract custom where its sales revenue may be judged against the size of the market where it is located, the competition it faces locally and so on.

Once the type of efficiency to be assessed is clarified potential input-output variables can be put forth, drawing in large measure on 'industry experience', that is on the experience of those working within the units being assessed or generally familiar with the detail of their operations. The input variables should capture all resources and the output variables all outcomes having a bearing on the type of efficiency being assessed. In addition, contextual factors impacting the transformation of inputs to outputs should also be reflected. Depending on the direction of the impact of a contextual factor it can be treated as an input or as output within the DEA model. In

---

[1] Strictly speaking exclusivity and exhaustiveness can be relaxed if we can assume that any omitted variables will not alter the proportionality across the DMUs of the values of the input-output variables being used.

some instances, especially when a prior assumption is not to be made that a contextual factor impacts the efficiency of transformation of inputs to outputs so called *two-stage* assessments are used. In the first stage contextual factors are ignored and a DEA assessment is carried out using non-contextual input-output variables. In the second stage a *Tobit* regression of the DEA scores on the contextual variables is carried out to identify those which impact the efficiency scores and adjust the latter appropriately. For more details on such two-stage approaches see Coelli et al. (1998) and Cooper et al. (2000). (Tobit regression is used because DEA scores are *censored* from above at 1.)

The initial set of potential input-output variables can be refined using a combination of statistical tests and/or sensitivity analysis. Statistical tests of association between proposed input-output variables can be carried out (e.g. correlations, analysis of variance, OLS regression). In particular, where a single input or alternatively a single output is involved regression may be used prior to running a DEA assessment to identify the factors most likely to fit the input-output correspondence being proposed. This is different from using regression to carry out the assessment itself since establishing the presence or absence of a mutual dependence of input-output variables is less demanding than establishing the precise formula of the dependence. Even so, caution is needed to always combine statistical results with industry experience in arriving at the final input-output set as statistical tests cannot in themselves prove the presence or absence of causality between input-output variables.

Sensitivity analysis may help in further refining a proposed input-output set. For example input or output variables which are deemed by those familiar with the working of the DMUs as having secondary role can be assessed for their impact on the results by running assessments with and without the variables concerned. Where only a few, potentially unusual, DMUs are impacted by such secondary variables the variables can be dropped from the assessment. (E.g. See Thanassoulis (2000a) for sensitivity analysis of this type used in the context of assessing UK water companies.) Otherwise, the variables will need to be retained in the assessment.

The ultimate aim is that the input-output set used should conform to the exclusivity, exhaustiveness and exogeneity requirements and should involve as few variables as possible.

The need that input-output variables should be as few as possible arises out of the need to retain discriminatory power on the comparative efficiencies of the units being assessed. For example it is quite clear that an additional input or output variable corresponds to an additional constraint within the envelopment model [M4.1]. So long as such a constraint is not redundant (by virtue of being linearly dependent on other constraints) it can only reduce the set of feasible solutions to that model. This in turn can only lead to a 'worsening' of the objective function value in the form of a larger efficiency rating for the DMU being assessed. Similarly, the fewer the DMUs being assessed the fewer the constraints within the value-based DEA model [M4.5]. This places fewer restrictions on the feasible set of solutions to that model. This in turn can only lead to a rise in the optimal objective function value (efficiency rating) yielded by that model. Thus, normally,

the larger the number of input and output variables used in relation to the number of DMUs being assessed the less discriminating on efficiencies the DEA assessment.

In conclusion:

- The input-output variables should be chosen in concordance with the type of efficiency being assessed.

- The input-output variables should conform to the exclusivity, exhaustiveness and exogeneity principles.

- The input-output variables should be as few as possible.

- The more the DMUs being assessed the more the input-output variables which can be tolerated without loss of discriminatory power on efficiencies.

5.3   INFORMATION OBTAINED IN THE COURSE OF A DEA ASSESSMENT

We focus here on some of the most common uses made of the information that flows directly from the solution of a DEA model. We limit ourselves to the DEA models we have covered so far. We shall introduce in later chapters modifications to these models and discuss their uses as the modified models are presented. The models covered so far are used most commonly for the following:

- Ascertaining a measure of the efficiency of each DMU;

- Getting a view on the robustness of its efficiency measure;

- Where the DMU is Pareto-efficient:
  - Identifying the areas in which it might prove an example of good practice for other DMUs to emulate;
  - Getting a view on the scope for the DMU to be a role-model for other DMUs.

- Where the DMU is Pareto-inefficient:
  - Identifying efficient DMUs whose operating practices it may attempt to emulate to improve its performance;
  - Estimating target input-output levels which the DMU should in principle be capable of attaining under efficient operation.

Any one of the DEA models we have covered so far will provide the information listed above. However, this requires recourse to primal-dual relationships of linear programming models which we have not covered in this book. We shall instead, therefore, solve an envelopment and a value-based DEA model separately and illustrate the information which can be derived in each case directly and without explicit recourse to primal-dual relationships. Later we shall introduce DEA software and demonstrate how it too can be used to derive the type of information listed above.

5.4   INTERPRETING THE SOLUTION OF A DEA ENVELOPMENT MODEL

The information most readily obtainable from a DEA envelopment model is the following:

a) Ascertaining a measure of the efficiency of the DMU;
b) Where the DMU is Pareto-efficient:
   Getting a view on the scope for the DMU to be a role-model for other DMUs.
c) Where the DMU is Pareto-inefficient:
   - Identifying efficient DMUs whose operating practices it may attempt to emulate to improve its performance;
   - Estimating target input-output levels which the DMU should in principle be capable of attaining under efficient operation.

We shall look first at the information the model yields in respect of Pareto-inefficient and then in respect of Pareto-efficient DMUs.

5.4.1 Pareto-Inefficient DMUs

Let us use as a vehicle the input minimising envelopment DEA model [M4.1] in respect of Centre 1, Example 4.1 in Chapter 4. The model to be solved is [M4.2] (Chapter 4) which is reproduced below as model [M5.1]. (The Greek λ's in [M4.2] have been replaced by L in [M5.1].)

Min	100k-0.001SL-0.001SC-0.001SD-0.001SP	[M5.1]

Subject to:

Labour:	4.1L1 + 3.8L2 + 4.4L3 + 3.2L4 + 3.4 L5-4.1k + SL = 0
Capital:	2.3L1 + 2.4L2 + 2.0L3 +1.8L4 + 3.4 L5-2.3k + SC =0
Distance:	4.3L1 + 3.9L2 + 4.1L3 +5.2L4 + 4.2 L5 - SD = 4.3
Packages:	90L1 + 102L2 + 96L3 + 110L4 + 120 L5 - SP = 90
	L1, L2, L3, L4, L5, SL, SC, SD, SP ≥ 0, k free.

In [M4.1] the coefficient of k was 1 but we are using here 100. The choice of 100 (and 0.001 for ε in [M4.1]) is arbitrary. The only requirement is that the coefficient of k in [M4.1] be much much larger than ε so that pre-emptive priority is given to the minimisation of k. (See also Chapter 4, Model [M4.1] where it was noted that in practice (but not here) no real value for ε is used and the model is solved by a two-phase optimisation process.)

Using Lindo as a solver (at the time of writing a free trial version of Lindo could be downloaded from http://www.lindo.com) the following results are obtained in respect of [M5.1]:

VARIABLE	VALUE
k	0.647157
SL	0.007191
SC	0.000000
SD	0.000000
SP	0.961538
L1	0.000000
L2	0.000000
L3	0.000000
L4	0.826923
L5	0.000000

The information we can glean from this solution is:

Measure of the efficiency of Centre 1:

The value of k at the optimal solution to [M5.1] is 0.647 (to three decimal places). This is what we defined in Chapter 2 as the *Technical Input Efficiency* of Centre 1. The value of 0.647 reflects the maximum radial contraction possible to the input levels of Centre 1 without detriment to its output levels. Thus Centre 1 is Pareto-inefficient.

Identifying efficient DMUs whose operating practices Centre 1 may attempt to emulate to improve its performance.

Such DMUs are the efficient peers or referents to Centre 1. These were identified (see Chapter 4) by the fact that their corresponding L values are positive at the optimal solution to [M5.1]. There is only one such positive value here, namely L4 = 0.827. Thus the efficient peer Centre 1 can attempt to emulate is Centre 4. As noted in Chapter 4, the efficient peers have a similar input-output mix to that of the inefficient DMU. This is indeed the case here. Of the 5 centres in Table 4.1, Centres 4 and 5 are Pareto-efficient. We can see quite clearly why it is Centre 4 rather than 5 that is being used as an efficient peer for Centre 1 if we look at the ratios the input levels are to each other and do the same for the output levels. The results are as follows:

Centre	Labour/Capital	Distance/Packages
1	1.783	0.0478
4	1.778	0.0473
5	1	0.035

Clearly the labour to capital ratio of Centre 1 at 1.783 is much closer to the corresponding ratio of 1.778 of Centre 4 than that of Centre 5. Similarly the ratio of Distance to Packages of Centre 1 is much closer to the corresponding ratio of Centre 4 than to that of Centre 5.

In practical applications non-specialists find the identification of efficient peers especially useful for gaining an intuitive feeling for the comparative efficiency results yielded by DEA. This is done by a straightforward comparison of the inefficient DMU with its peers. For example we can see by inspection of the data in Table 4.1 that while Centre 4 has lower labour and capital levels than Centre 1, it has larger levels of distance covered and packages delivered. Thus, it is clear even without recourse to DEA, that Centre 4 is operating more efficiently than Centre 1. Moreover, in assessments under constant returns to scale we may be able to scale up or down one or more efficient peers to see the basis underlying the efficiency rating of the inefficient DMU. For example here if we scale Centre 4 by 0.827 and contrast the results with Centre 1 we have the following picture:

		Labour	Capital	Distance	Packages
Centre 1 (C1)		4.1	2.3	4.3	90
Centre 4 (C4) × 0.827	=	2.65	1.489	4.3	90.97
(C4 × 0.827) / C1	=	0.6454	0.6472	1	1.01

As can be seen in the last row above, the scaled down by 0.827 Centre 4 offers the same distance covered as Centre 1 and slightly more packages delivered but it uses no more than 64.72% of the inputs Centre 1 uses and this underlies the Pareto-efficiency rating of 64.72% noted above. In this simple case of a single efficient peer the scaled peer is nothing other than the DEA targets defined in (4.1) in Chapter 4. Thus the ratios computed above will yield the precise DEA efficiency rating. However, even when we have multiple efficient peers the scaling of one or more peers can yield a fair approximation to the efficiency rating of the inefficient DMU as will be seen in Section 6 when we re-visit efficient peers as yielded by DEA software.

Estimating target input-output levels which Centre 1 should in principle be capable of attaining under efficient operation

Such targets in the case of the input minimising envelopment model were defined for the general case in Chapter 4, in expression (4.1). That expression, in the context of the solution to model [M5.1] is as follows:

$$\text{Labour}^t = 0.827\,\text{Labour}_4 = 0.827 \times 3.2 = 2.65$$
$$\text{Capital}^t = 0.827\,\text{Capital}_4 = 0.827 \times 1.8 = 1.49$$
$$\text{Distance}^t = 0.827\,\text{Distance}_4 = 0.827 \times 5.2 = 4.3$$
$$\text{Packages}^t = 0.827\,\text{Labour}_4 = 0.827 \times 110 = 90.97.$$

The superscript t indicates target level, the subscript 4 indicates Centre 4 and 0.827 is the only positive L value, namely L4, at the optimal solution to model [M5.1]. Thus we conclude that under efficient operation, Centre 1 should be capable of covering a distance of 4.3 thousand km and deliver 90.97 thousand packages using 2,650 hours of labour and $1.49m capital. These target levels are of course compatible with the operational interpretation of the efficiency rating of 64.71% of Centre 1. That interpretation was that the Centre can contract radially its inputs to at least 64.71% of their observed levels without detriment to its output levels. (The contracted input levels are 0.6471 × (4.1, 2.3) = (2.78, 1.49) so they are compatible with the targets above. However, the target levels above indicate that following the radial contraction of the input levels, labour can contract further from the level of 2.78 to the level of 2.65 thousand hours and packages delivered can rise by 970.)

5.4.2 Pareto-Efficient DMUs

The envelopment model gives information on the extent to which a Pareto-efficient DMU is used as an efficient peer (benchmark) for other DMUs. Such information helps us gauge how similar are the operating practices and environment of the Pareto-efficient DMU in question to those of other less efficient DMUs. In order to illustrate the use of this type of information in DEA assessments let us reconsider the Distribution Centres of Table 4.1 of Chapter 4. If we solve the input minimising DEA model in [M4.1] in respect of each centre in turn and record the efficient peers to each inefficient DMU and the associated λ values the following picture emerges:

Centre	Peers and associated λ-values	
	Centre 4	**Centre 5**
1	0.827	
2	0.778	0.137
3	0.873	
4	1	
5		1

Thus there are three Pareto-inefficient centres (1-3), and two Pareto-efficient centres, (4 and 5). Focusing on the Pareto-efficient centres two aspects are of interest: *How frequently is each Pareto-efficient centre used as an efficient peer* and *how strong is its influence on the targets estimated for inefficient centres?*

Frequency of use as an efficient peer
Centre 4 is used as an efficient peer to all Pareto-inefficient centres so its frequency of use as an efficient peer, expressed as a percentage of the number of Pareto-inefficient DMUs, is 100%. Centre 5 is an efficient peer to only one Pareto-inefficient centre and so its corresponding frequency is 33.33%. The relative frequency of use of a Pareto-efficient DMU as a peer has two practical uses. Firstly, we have enhanced confidence that a DMU which is a frequent efficient peer is genuinely a well performing DMU because it outperforms many DMUs. Secondly, such a DMU is likely to be a better role model for less efficient DMUs to emulate because its operating practices and environment match more closely those of the bulk of DMUs than is the case for a Pareto efficient DMU which is rarely an efficient peer. (See Chapter 4 for the argument that efficient peers have a similar mix of input-output levels to the inefficient DMU concerned.) Thus, in the foregoing example Centre 4 more than Centre 5 is genuinely efficient and more suitable to use as a role model to be emulated by other centres. These conclusions based on the frequency with which a Pareto-efficient DMU is

used as a peer need to be tested against the contribution the DMU makes to targets of inefficient DMUs as outlined next.

Influence on the targets estimated for inefficient DMUs

We would expect that a DMU featuring frequently as an efficient peer to inefficient DMUs will also have a great impact on the targets estimated for inefficient DMUs. This, however, need not always be so and we can get a better measure of the influence an efficient DMU exerts on the targets estimated for inefficient DMUs by direct computation of its contribution to such targets. (See (4.1) in Chapter 4 for the summation of the contributions of efficient peers to derive the target levels of each Pareto-inefficient DMU.) Thus, in the case of Centres 1-3 which are inefficient we have the following picture:

	Labour	**Capital**	**Distance**	**Packages**
Sum of target levels (Centre 1, 2 and 3)	8.39	4.926	13.461	289.02
Percentage contributed by Centre 4	94.45	90.544	95.725	94.311

Centre 4 contributes of the order of over 90% of the target levels of inefficient DMUs and it is an efficient peer for 100% of them. (Here the contribution made by Centre 4 to the targets of Pareto-inefficient DMUs supports the conclusions reached above that, it operates in terms of environment and practices more in line with the rest of the centres than is the case for Centre 5. Had the contribution to targets made by Centre 4 been low we could not safely draw the conclusion that it operates in terms of environment and practices in line with the bulk of the centres. It must, be noted in passing though that it is in principle possible that a Pareto-efficient DMU which is an efficient peer for few, if any, DMUs to be very productive but simply operating in an environment and/or using practices which are very dissimilar to those of the rest of the DMUs being assessed.)

5.5 INTERPRETING THE SOLUTION OF A VALUE-BASED DEA MODEL

Value-based DEA models are most suitable for obtaining the following type of information:
a) Ascertaining an overall measure of the efficiency of each DMU;
b) Getting a view on the robustness of the efficiency measure of a DMU;
c) Where the DMU is Pareto-efficient:
    Identifying the areas in which it might prove an example of good operating practice for other DMUs to emulate;

d) Where the DMU is Pareto-inefficient:
    Identifying Pareto-efficient DMUs whose operating practices it may attempt to emulate to improve its performance.

5.5.1 Pareto-Efficient DMUs

The value-based DEA model in [M5.2] is the instance of model [M4.5] in respect of Centre 4 of Example 4.1.

Max	$Z_1 = 5.2u_D + 110u_P$	[M5.2]
Subject to:		
Normalisation:	$3.2v_L + 1.8v_C = 100$	
Centre 1:	$4.3u_D + 90u_P - 4.1v_L - 2.3v_C \leq 0$	
Centre 2:	$3.9u_D + 102u_P - 3.8v_L - 2.4v_C \leq 0$	
Centre 3:	$4.1u_D + 96u_P - 4.4v_L - 2v_C \leq 0$	
Centre 4:	$5.2u_D + 110u_P - 3.2v_L - 1.8v_C \leq 0$	
Centre 5:	$4.2u_D + 120u_P - 3.4v_L - 3.4v_C \leq 0$	
	$u_D, u_P, v_L, v_C \geq 0.001.$	

We have set the normalisation constant to 100, primarily to reduce round-off errors in the computed values of the DEA weights u_D, u_P, v_L and v_C. (See Section 7 for an elaboration on this point.) The weights have been restricted not to exceed the small positive value of 0.001. We have chosen arbitrarily this value as being 'very small and much below 1' to represent the non-Archimedean infinitesimal ε found in the generic DEA model underlying [M5.2], namely model [M4.5]. This enables us to solve a single linear programming model here to illustrate the concepts involved without extensive recourse to duality. In practice, as noted earlier, no exact value for ε is used and the envelopment model corresponding to [M5.2] is solved in two phases. (See the related comment on model [M4.1] in Chapter 4.)

The solution yielded by Lindo in respect of model [M5.2] is as follows.

OBJECTIVE FUNCTION VALUE = 100.000

VARIABLE	VALUE
u_D	1.928
u_P	0.818
v_L	31.249
v_C	0.001

The information this solution yields is interpreted as follows:

Ascertaining a measure of the efficiency of Centre 4
   The objective value of 100 is the measure of the efficiency of the centre which is therefore Pareto-efficient.

Getting a view of the robustness of the efficiency of Centre 4
   As we saw, the **envelopment model** indirectly can enhance our confidence about the efficiency measure of a Pareto-efficient DMU when the DMU is found to have been an efficient peer to many inefficient DMUs and to have had a large impact on their targets. **Value-based DEA models** yield more direct information about the robustness of the efficiency measure of a DMU by means of the *virtual* input and output levels of the DMU being assessed. It is recalled that in the context of the generic value-based DEA model [M4.5] the ith virtual input of DMU j_0 is $x_{ij_0} v_i^*$ and its rth virtual output is $y_{rj_0} u_r^*$. The superscript * denotes the optimal value of the weight concerned in model [M4.5].

   The model [M5.2] yields the following virtual input and output levels in respect of Centre 4:

	Optimal DEA weight	Input, Output level	Virtual Input (Output)
Distance	1.928	5.2	10
Packages	0.818	110	90
Labour	31.249	3.2	100
Capital	0.001	1.8	0

   The virtual input and output levels reflect the extent to which the efficiency rating of a Pareto-efficient DMU is underscored by each one of its input and output levels. Thus Centre 4 relies for its efficiency rating essentially on delivering a large number of packages for the labour it uses. This is deduced from the fact that of the total virtual output of 100 ('units'), 90 ('units') are accounted for by the level of packages delivered. The entire virtual input of 100 is accounted for by its labour.

   The robustness of the efficiency rating of a DMU can be judged by the extent to which the DMU relies on a limited number of inputs and outputs to be rated Pareto-efficient. In an extreme case, a DMU may have its entire virtual input accounted for by one input and its entire virtual output accounted for by one output. Such a DMU relies on the ratio of a single output to input to be deemed efficient and so it cannot be said to have a very robust efficiency rating. We do not know whether it will continue to be deemed to have good performance relative to other DMUs if more inputs and outputs are to be reflected in the efficiency rating. (The closer a DMU is to this extreme case of its total virtual input and output being accounted for by a small subset of the inputs and outputs respectively the less robust its efficiency rating.) It is a subjective judgement as to what close means in this context. For example Centre 4 above on the face of it does not have robust efficiency rating.

It is very important, however, to note in the context of the foregoing discussion that in practice we almost always have multiple optimal solutions to model [M4.5] when it identifies a Pareto-efficient DMU. Thus, a low virtual level on an input or an output in one of the optimal solutions does not necessarily mean that the DMU cannot have a high virtual level on that input or output in some other optimal solution to the model. This in turn means that we cannot really take a view that the efficiency rating of a DMU is not robust until we explore the range of weights which give the DMU its maximum efficiency rating. This is cumbersome to do but there is one simple way in which we can make some progress in this context. We can solve [M4.5] modified as follows:

Max $\qquad$ p $\hfill$ [M5.3]

Subject to:

$$\sum_{r=1}^{s} u_r y_{rj_0} = h^*_{j_0}$$

$$\sum_{i=1}^{m} v_i x_{ij_0} = 1$$

$$\sum_{r=1}^{s} u_r y_{rj} - \sum_{i=1}^{m} v_i x_{ij} \leq 0 \quad j = 1...j_0...N$$

$$
\begin{aligned}
u_r \, y_{rj_0} &\geq p & r &= 1...s \\
v_i \, x_{ij_0} &\geq p & i &= 1...m \\
u_r &\geq \varepsilon & r &= 1...s \\
v_i &\geq \varepsilon & i &= 1...m.
\end{aligned}
$$

Where $h^*_{j_0}$ is the efficiency rating of DMU j_0 obtained by solving [M4.5]. Notation in [M5.3] is otherwise as in [M4.5]. Model [M5.3] determines the maximum level of the lowest virtual input or output level of DMU j_0 compatible with its DEA efficiency rating. Solving [M5.3] in respect of Centre 4 leads to a solution in which the virtual input-output levels are as follows:

	Optimal DEA weight	Input, Output level	Virtual Input (Output)
Distance	9.615	5.2	50
Packages	0.4545	110	50
Labour	15.625	3.2	50
Capital	27.7778	1.8	50

Clearly Centre 4 can attain its efficiency rating of 100% even when all inputs and outputs contribute equal amounts to total virtual input and output and so we have <u>no</u> evidence that its efficiency rating is not robust.

Identifying areas in which a DMU might offer good operating practices for other DMUs to emulate.

Operating practices which are conducive to efficiency would naturally be sought at Pareto-efficient DMUs. However, the imputation of DMU-specific marginal input-output values makes it necessary to proceed with caution on this issue. A DMU may appear Pareto-efficient through imputing low enough values to certain inputs and/or outputs so as to effectively ignore them.

Pareto-efficient DMUs are likely to offer good operating practices on the inputs and outputs to which they have large virtual levels.

A high virtual level on a particular output does not necessarily mean the output is the result of efficient operating practices as there are many possible trade offs between outputs and between inputs. Rather, a high virtual input or output is an indication that the corresponding operating practices may be efficient but this needs to be investigated 'on the ground' at DMU level, outside the DEA context.

Two additional points which should be noted so far as the identification and dissemination of efficient operating practices is concerned are the following. Firstly, good operating practices on uncontrollable factors should be disseminated the same as those relating to controllable ones. The DMU may have control over how it operates with its given levels of uncontrollable variables. Secondly, efficient practices can be disseminated not only from Pareto-efficient to Pareto-inefficient but also from one to another Pareto-efficient DMU when the two offer efficient operating practices in complementary areas. DEA assesses <u>relative</u> rather than <u>absolute</u> performance and so a Pareto-efficient DMU may always be capable of improving its performance further. It is repeated, however, that as noted above, in practice the DEA model [M4.5] will very often have multiple optimal solutions when it identifies a Pareto-efficient DMU. Thus, a low virtual level for an input or an output in one optimal solution does not necessarily imply the DMU has poor operating practices on that area.

5.5.2 Pareto-Inefficient DMUs

Model [M5.4] is the instance of [M4.5] in respect of Centre 1 of Example 4.1 in Chapter 4.

Max $\qquad$ $Z_1 = 4.3u_D + 90u_P$ $\qquad\qquad$ [M5.4]

Subject to:

Normalisation: $\quad 4.1v_L + 2.3v_C = 100$

Centre 1: $\quad 4.3u_D + 90u_P - 4.1v_1 - 2.3v_C \leq 0$

Centre 2: $\quad 3.9u_D + 102u_P - 3.8v_1 - 2.4v_C \leq 0$

Centre 3: $\quad 4.1u_D + 96u_P - 4.4v_1 - 2v_C \leq 0$

Centre 4: $\quad 5.2u_D + 110u_P - 3.2v_1 - 1.8v_C \leq 0$

Centre 5: $\quad 4.2u_D + 120u_P - 3.4v_1 - 3.4v_C \leq 0$

$\qquad u_D, u_P, v_1, v_C \geq 0.001.$

As in model [M5.2] we have set the normalisation constant to 100, and the weights have been arbitrarily restricted not to exceed the small positive value of 0.001. The solution yielded by Lindo in respect of model [M5.4] is as follows.

OBJECTIVE FUNCTION VALUE = 64.71475

VARIABLE	VALUE
u_D	15.029
u_P	0.001
v_L	0.001
v_C	43.476

The solution in this case reveals similar information to that detailed above for the case of a Pareto-efficient DMU.

Ascertaining a measure of the efficiency of Centre 1

The objective function value of 64.714 is the measure (in percentage terms) of the technical input efficiency of the centre. Centre 1 is therefore Pareto-inefficient. It should be capable of lowering its input levels to no more than 64.714% of their observed levels without any detriment to its output levels. This is of course the same result as we obtained earlier by solving model [M5.1].

Getting a view of the robustness of the efficiency of Centre 1

We can compute the virtual input and output levels of the centre to ascertain what underlies its efficiency rating. The results are as follows:

	Optimal DEA weight	Input, Output level	Virtual Input (Output)
Distance	15.029	4.3	64.624
Packages	0.001	90	0
Labour	0.001	4.1	0
Capital	43.476	2.3	100

The entire virtual input is accounted for by capital while the entire virtual output is accounted for by the distance covered by the Centre. This suggests the efficiency rating of 64.714% is not very robust as packages delivered and labour used are ignored. However, there could be alternative optimal solutions to [M5.4] in which we have different virtual input and output levels. Solving the equivalent to model [M5.3] but in respect of Centre 1 leads to the following virtual levels.

	Optimal DEA weight	Input, Output level	Virtual Input (Output)
Distance	15	4.3	64.5
Packages	0.0024	90	0.216
Labour	0.0533	4.1	0.22
Capital	43.383	2.3	99.78

Thus it is indeed the case that Centre 1 cannot attain its maximum efficiency rating of about 64.7% if it is to take into account packages delivered and labour used in addition to distance covered and capital used. So the subjective view is that the efficiency rating of the Centre is not very robust.

Identifying efficient DMUs whose operating practices Centre 1 may attempt to emulate to improve its performance

What is called for here is the identification of the efficient peers to Centre 1. These are centres whose corresponding constraints are binding at the optimal solution to [M5.4]. At the optimal solution to [M5.4] the constraint $5.2u_D + 110u_P - 3.2v_L - 1.8v_C \leq 0$, corresponding to Centre 4, is binding as can be readily deduced from either one of the two sets of optimal DEA weights given above; (e.g. $u_D = 15.029$, $v_{2C} = 43.476$, $u_P = v_L = 0$). No other constraint corresponding to a centre is binding at the optimal solution to [M5.4]. Thus Centre 4 is the sole efficient peer to Centre 1. This suggests Centre 4 is the 'nearest' we can find to an efficient Centre which has a similar input-output mix to Centre 1 and so the two may be operating in similar environmental conditions. This would make Centre 4 the nearest we can find to an efficient unit whose operating practices Centre 1 should attempt to emulate.

5.6 AN ILLUSTRATIVE USE OF *WARWICK DEA SOFTWARE*

The information obtained from the DEA models discussed in sections 5.2 and 5.3 can be obtained much more conveniently (and quickly) using DEA software. We shall illustrate the use of the *Warwick DEA Software*. A limited version of the software, capable of assessing only 10 DMUs accompanies this book. A summary of the capabilities of the current version of the software appears in Appendix 5.1. A limited User Guide on the software can be found in Chapter 10.

The software is upgraded periodically to incorporate additional features reflecting new research developments in the field. Interested readers can contact the author on dea.et@btinternet.com for up to date information on the software, its capabilities and how to acquire fuller or more up to date versions of it.

We shall use a small bank which has 5 branches as a vehicle for illustrating the software. The input-output data for the branches appears in Table 5.1 and it is a subset of the data included with the software accompanying this book. We will use the *Warwick DEA Software* to:

i)   Construct a table showing the relative efficiencies of the branches;
ii)   Calculate the DEA weights and the virtual-input and output
        levels of a sample of branches;
iii) Identify the efficient peers to the least efficient branch;
iv)   Estimate input minimising targets for the least efficient branch.

You may like to consult at this point the User Guide in Chapter 10 before reproducing the computations in this section.

Table 5.1.   Data on Bank Branches

Branch	Staff Costs	Other Costs	Deposit Accounts	Loan Accounts	New Deposit Accounts
	($000)	($000)	(000)	(00)	(00)
1	143	62	28	7	3
2	183	63	52	12	15
3	515	250	120	80	27
4	178	79	82	15	3
5	189	73	52	32	4

A file is prepared first containing the input-output data of Table 5.1 in a simple, pre-specified, format, illustrated in Figure 5.1. The file must be

ASCII i.e. saved as *text only*. (See Chapter 10 for more on the input file format.)

```
-StaffCost –OtherCost +DepAccs +LoanAcc +NewDAcc
   B1 143 62 28 7 3
   B2 183 63 52 12 15
   B3 515 250 120 80 27
   B4 178 79 82 15 3
   B5 189 73 52 32 4
   END
```

Figure 5.1.   Illustrative Input File to *Warwick DEA Software*

This file is read into the DEA software at run time using the *File* menu, illustrated in Figure 5.2.

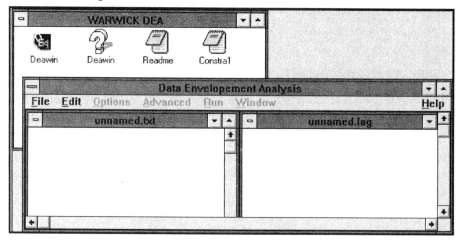

Figure 5.2.   The *Warwick DEA Software* Screen Before Data Has Been Read in

Quite evidently the input-output variables in this case are as follows:

Inputs
- Staff costs(STAFFCST)
- Other costs (OTHERCST)

Outputs
+ Deposit accounts (DEPAC)
+ Loan accounts(LOANS)
+ New deposit accounts (NEWAC)

Operating expenditure could be aggregated to give a better measure of the cost efficiency of a branch. (We have no input prices to estimate allocative efficiency here.) Using disaggregated expenditure as suggested above is appropriate when the branch management being assessed has no control over the relative mix of staff and other costs of their branch.

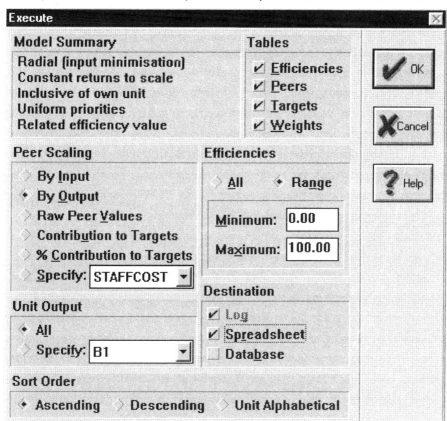

	STAFFCOST	OTHERCOST	DEPACCS	LOANACC	NEWDACC
B1	143.00	62.00	28.00	7.00	3.00
B2	183.00	63.00	52.00	12.00	15.00
B3	515.00	250.00	120.00	80.00	27.00
B4	178.00	79.00	82.00	15.00	3.00
B5	189.00	73.00	52.00	32.00	4.00

Figure 5.3.   *Warwick DEA Software* Menu Options after Data Has Been Read in

Figure 5.4.   The *Execute* screen for Specifying Output Content and Format from the *Warwick DEA Software*

Figure 5.3 shows a screen of the *Warwick DEA software* after the input-output data has been read in. The drop-down menus of the software are covered in Chapter 10. In brief, the *Options* menu enables the user to specify whether input or output efficiency is to be estimated, whether *constant* or *variable* returns to scale are to be assumed, whether the unit being assessed is to feature within potential referent units (peers) and so on. (Variable returns to scale are covered in Chapter 6.) The *Advanced* menu makes it possible to specify weights restrictions within the value-based DEA model [M4.5] (or [M4.6] (see also Chapter 8).) The command *Execute* in the *Run* menu enables the user to specify the results to be printed and the format of the results (spreadsheet readable file or log file or both) and so on. See Figure 5.4.

The Tables shown in the remainder of this section are taken from the *Warwick DEA Software* log file. The results can also be printed into a file for importation into a spreadsheet for further manipulation. Note that total Virtual Input (Model [M4.5]) is normalised to 1 in the software, which uses two-phase optimisation in place of an ε in the context of the DEA models covered so far. (See Chapter 4.) Further notation is as follows:

– IO = Input/Output,
– Inputs are preceded by "-" and outputs by "+".

Table of efficiencies

B1	52.51	B3	100	B5	100
B2	100	B4	100		

In cases of multiple optimal solutions to the value-based DEA model, the DEA weights printed are those resulting from model [M5.3] which maximise the lowest virtual input or output level compatible with the DMU's DEA efficiency rating. E.g. in the case Branch B2 we have:

Virtual IOs for B2-efficiency 100%

VARIABLE	VIRTUAL IOs	IO WEIGHTS
-OTHERCST	72.63%	0.01153
-STAFFCST	27.37%	0.00150
+DEPAC	27.37%	0.00526
+LOANS	27.37%	0.02281
+NEWAC	45.26%	0.03017

Branch B2 is relatively efficient in the main due to the number of new accounts it has relative to its "other costs". Looking at the tables of DEA weights relating to Branches 4 and 5 we note that Branch 5 is efficient when a single output (loans) is given prominence. Branches 4 and 5 may benefit by exchanging operating practices as they each appear strong in a different field of activities.

Virtual IOs for B4-efficiency 100.00%		
VARIABLE	**VIRTUAL IOs**	**IO WEIGHTS**
-OTHERCST	8.27%	0.00105
-STAFFCST	91.73%	0.00515
+DEPAC	83.46%	0.01018
+LOANS	8.27%	0.00551
+NEWAC	8.27%	0.02757

Virtual IOs for B5 efficiency 100%		
VARIABLE	**VIRTUAL IOS**	**IO WEIGHTS**
-OTHERCST	46.74%	0.00640
-STAFFCST	53.26%	0.00282
+DEPAC	13.60%	0.00262
+LOANS	72.80%	0.02275
+NEWAC	13.60%	0.03400

We would need to identify all optimal sets of DEA weights for each branch to be certain the branches are strong only in the areas highlighted in the tables reproduced here.

The efficient peers to a branch can help demonstrate what underlies its DEA efficiency rating. For example Branch B2, an efficient peer to Branch B1, has almost twice (or more) the output levels of B1 yet its input levels are never more than 30% higher than those of B1. See the Table of unscaled peers to B1 reproduced here. The Lambda's relate to those in Model [M4.1].

(Unscaled) Peers for B1 - efficiency 52.51%				
B1		**B2**	**B3**	**B4**
ACTUAL	LAMBDA	0.103	0.028	0.236
62.0	-OTHERCST	63.0	250.0	79.0
143.0	-STAFFCST	183.0	515.0	178.0
28.0	+DEPAC	52.0	120.0	82.0
7.0	+LOANS	12.0	80.0	15.0
3.0	+NEWAC	15.0	27.0	3.0

We can take the comparison of an inefficient DMU with its efficient peers further. Take for example Branch B1. We can specify within the *Warwick DEA Software* that '*peers are to be scaled according to outputs*' as illustrated in Figure 5.4. This will cause the software to scale the input-output levels of each efficient peer of Branch B1 so that at least one output level of the peer equals that of Branch B1 and no output level of the peer is lower than the corresponding output level of Branch B1. Table 5.2 shows the peers to Branch B1 after they have been scaled in this way. The scaling constant used in the case of each peer is also shown.

Table 5.2.   Peers to Branch B1 Scaled by Output

B1		B2	B3	B4
ACTUAL	LAMBDA	0.103	0.028	0.236
	SCALE	0.583	0.233	1.000
62.0	-OTHERCST	36.8	58.3	79.0
143.0	-STAFFCST	106.8	120.2	178.0
28.0	+DEPAC	30.3	28.0	82.0
7.0	+LOANS	7.0	18.7	15.0
3.0	+NEWAC	8.7	6.3	3.0

The scaled efficient peers can be used to demonstrate in simple terms how the performance of the inefficient branch is poorer and even lead to a rough estimate of the efficiency rating of the inefficient branch. For example we can see in Table 5.2 that Branch B2 has been scaled to 58.3% of its original input-output levels. The scaled output levels of Branch B2 are never below the corresponding output levels of Branch B1 yet its two input levels are 59.35% and 74.68% of those of Branch B1. This not only shows that the performance of Branch B1 is poorer than that of Branch B2, but it also shows that relative to Branch B2 Branch B1 must be able to lower its costs to no more than 74.68% of their original levels which means the input efficiency of the branch cannot be in excess of 74.68%. (We know the input efficiency of Branch B1 to be 52.51% but the single scaled peer we have used <u>cannot</u> give a good estimate of this when its scaled outputs exceed so much those of Branch B1 as is the case in Table 5.2.)

Turning to targets which would render a unit efficient, if output levels are exogenous, as arguably is the case with the bank branches, then the input minimising model would be more appropriate to use to estimate targets. The results are as follow:

Targets for B1-efficiency 52.51%

VARIABLE	ACTUAL	**TARGET**	TO GAIN	ACHIEVED
-OTHERCST	62.0	**32.1**	48.3%	51.7%
-STAFFCST	143.0	**75.1**	47.5%	52.5%
+DEPAC	28.0	**28.0**	0.0%	100.0%
+LOANS	7.0	**7.0**	0.0%	100.0%
+NEWAC	3.0	**3.0**	0.0%	100.0%

(The targets are in bold.)

Thus, once the input-output data is ready the DEA software can readily produce much of the 'standard' information on the performance of the units being assessed. Further manipulation of the results can make new insights possible, most notably in the form of measures of productivity change over time or decompositions of inefficiencies into those attributable to policy decisions and managerial actions as elaborated in Chapter 7.

5.7   PRACTICAL TIPS FOR CARRYING OUT DEA ASSESSMENTS

This section gives a few tips in carrying out DEA assessments which would be helpful in practice.

Data scaling

Data should normally be scaled so that input-output levels do not take excessively large values, such as say over 1000. This is in order to reduce round-off errors in solving the DEA models involved. The DEA weights in [M4.5] or [M4.6] are inversely proportional to the levels of the inputs and outputs and so they will take very low or even negligible numerical values when the inputs and/or the outputs take very large values. This can lead to round-off errors. Apart from scaling the input-output levels it is also advisable to set the RHS of the normalisation constraint to a value larger than 1, say 100. This further scales up the DEA weights in [M4.5] and [M4.6]. (Note that the *Warwick DEA Software* automatically scales the input-output data to reduce round-off errors during the solution of the model. The data is scaled back to the user-supplied levels for the purposes of reporting the results once the relevant models have been solved.)

Lack of isotonicity

We shall discuss here non-isotonic outputs since the treatment of non-isotonic inputs is analogous.

Frequently a factor is met which though appearing as an output in the sense of being an outcome of the input to output transformation process being modelled, it is the case that, all else being equal, the lower its value the more efficient the unit. An example of such an output is the 'number of complaints received' when assessing service provider units or the amount of polluting effluent resulting from a manufacturing process. Wherever possible such outputs should be replaced with variables which are meaningful in the context of the assessment but which have *isotonic* properties in the sense of larger output levels reflecting better efficiency, all else being equal. In the example above, one may use as an output variable the number of service occasions which did not lead to a complaint instead of number of complaints received. (An alternative often employed is to subtract a non-isotonic output from a large positive number or divide it into a number so that the larger the non-isotonic output the lower the transformed output level used within the model. The results obtained are affected by the choice of arbitrary reference number used and so it is preferable to use a meaningful isotonic output wherever possible. See also Cooper et al. (2000) section 4.3 for alternative DEA models which are *translation invariant* in the sense that adding a constant to input or output levels will not impact the results.)

Zero input-output levels

Two issues arise in this case. Firstly not all the DMUs may be comparable in the sense that they are not all operating the same technology or face the same options of transforming inputs to outputs. Evidently DMUs which have zero on some input can secure the outputs being used in the assessment without using at all the inputs. Such DMUs differ from those which need at least some amount of the input concerned in order to operate at all. (E.g. a perinatal care unit which by law must have midwifery input may not be comparable to one which can opt not to engage midwives.) Similarly DMUs which have zero level on some output may not have certain input to output transformation options at all. It is, however, possible in principle for all DMUs to operate the same technology and simply for some of them to have negligible levels on certain inputs and outputs.

In such a case the second problem is that the assessment is likely to show DMUs having zero input levels artificially more efficient than they really are. This is because DMUs which have positive levels on the input that a DMU has zero level cannot be its efficient peers and so the comparative set of units is effectively reduced for DMUs with zero input levels. (To see this note that in the envelopment models [M4.1] and [M4.3] no λ or α respectively can be positive if the corresponding input is positive within a constraint that has a zero input in the RHS.) No similar problem arises with zero output levels. (In practice some users add a positive constant to the input and/or output levels to render them all positive. It must be stressed that this does affect the results obtained as the proportionality of input-output levels is altered by the value of the constant used, which is subjective. As noted above Cooper et al. (2000, section 4.3.2) present certain translation invariant DEA models.)

Data accuracy

DEA is a deterministic method and assumes in principle that all data is accurate. Inaccurate data of a DMU can have an impact depending on whether it renders incorrectly the DMU Pareto-efficient or Pareto-inefficient.

In the first instance, where the 'inaccurate DMU' has been inappropriately made Pareto-efficient the efficiencies of other DMUs for which it is an efficient peer may be underestimated. The opposite is true when the inaccurate DMU has been inappropriately deemed Pareto-inefficient. The DMUs which would have otherwise had it as their efficient peer may appear to have larger efficiency ratings than would be justified.

In all other cases, that is where the inaccurate data has not incorrectly affected the Pareto-efficiency status of the DMU concerned, the efficiencies of other DMUs are not affected by the inaccurate data at issue. Only the efficiency rating of the DMU concerned can be affected.

5.8   QUESTIONS

(Where the context permits the reader may employ the Warwick DEA Software accompanying this book to answer the questions below or to confirm answers derived by other means. See Chapter 10 for directions in using the limited version of the software accompanying this book.)

1.   The data below is to be used to assess the operating efficiencies of a set of warehouses belonging to a supermarket chain. *Stock* represents the value of goods in store. *Issues* are the order lots the warehouse packs ready for delivery to the supermarket's outlets, *Receipts* are the order lots it receives from suppliers and *Requisitions* are orders for goods it places with suppliers.

Table 5.3.   Warehouse Input-Output Data

	Stock ($m)	Wages ($00,000)	Issues (00)	Receipts (000)	Requisitions (000)
W1	4.05	6.75	54	74.25	40.5
W2	3.375	6.075	60.75	67.5	54
W3	5.4	8.1	74.25	60.75	40.5
W4	8.1	9.45	64.8	27	81
W5	3.105	4.725	37.8	67.5	33.75
W6	5.4	8.775	64.8	27	87.75
W7	9.45	13.5	108	87.75	76.95
W8	5.94	8.64	33.75	64.8	40.5
W9	4.05	6.75	60.75	86.4	56.7
W10	6.75	9.45	94.5	87.75	64.8
W11	6.75	9.45	60.75	87.75	54
W12	2.7	5.4	60.75	54	59.4
W13	6.75	9.45	87.75	33.75	47.25
W14	5.4	5.4	51.3	24.3	86.4
W15	2.7	4.05	27	67.5	20.25
W16	4.05	8.1	51.3	27	81
W17	9.45	14.85	91.8	86.4	72.9
W18	5.4	8.1	33.75	51.3	27
W19	4.05	5.4	60.75	90.45	43.2
W20	6.75	8.1	76.95	81	54

a)   Compute a DEA efficiency rating for each warehouse.

b) Comment on the relative performance of a Pareto-efficient warehouse. Your comments should cover contributory factors to its efficiency rating and the usefulness of such information in practice.

c) Comment on the relative performance of the least efficient warehouse. Your comments should cover:

♦ Warehouses which can be used as role models for the inefficient warehouse;

♦ Target levels that would render the warehouse efficient under the following scenarios:
- Activity levels at the warehouse are not expected to alter significantly in the near future;
- Additional activities are expected to be transferred from other warehouses to the least efficient one. How far can its activity levels rise before the latter will need additional resources?

2. Appendix 5.2. contains data on Metropolitan and London rates departments for 1982/3. The columns are as follows:
Costs=   The operating costs of the rates department.
Accounts= Number of rates accounts administered.
Rebates= Number of applications processed for rates rebates.
Summons= Number of summonses taken out to pursue late payers.
Value=   Net present value of rates charges collected.

a) Rank the departments in order of relative efficiency.
b) Comment on how confident you are of your results relating to relatively efficient departments. Could some of them be performing better than others? Are there areas in which some of them may be performing better than others? Explain your reasons.
c) Comment on the results relating to a least efficient department. Which is its group of efficient peers? To what extent do you find the peer group supports the efficiency rating of the inefficient department?
d) Estimate input minimising targets for the least efficient department. Repeat for output maximisation targets. Which set of targets would be the more sensible in practice? Why?
(*This question requires the use of DEA software to be carried out effectively.*)

3. The regional manager of a company supplying house builders wants to assess the relative efficiencies of the six stores serving building contractors in his region. The function of the stores is to receive

incoming goods from the company's factories and store them until they are required to supply a client builder. Stores are also responsible for the transport of the goods from the store to the builder. The manager collected the following data which relates to a three-month period.

Table 5.4.   Data on Building Materials Stores

Store	Equipment and Vehicles ($'00,000)	Staff Costs ($'00,000)	Volume of Goods Received to Store (100 tonnes)	Volume of Goods Transported to Clients (100 tonnes)
1	100	165	50	45
2	120	145	45	30
3	120	180	65	58
4	155	220	80	45
5	205	80	95	100
6	170	90	110	70

a) Assume the DEA efficiency of store 4 is 0.8. Interpret this efficiency rating in terms of the volume of goods the store should have been capable of handling in the three-month period concerned, without the need for additional resources.

b) Discuss briefly whether the variables used to assess the stores are adequate and indicate whether there are any additional variables you would deem appropriate to use, giving your reasons.

c) In order to assess one of the stores the manager set up an input minimisation DEA model in which two of the constraints were as follows:

$$120H-100L1-120L2-120L3-155L4-205L5-170L6 - SI1 = 0$$
$$45L1 + 30L2 + 58L3 + 45L4 + 100L5 + 70L6 - SO2 = 30$$

Complete the model.

d) At the optimal solution to the model in c) the slack variables in the output constraints are zero as are all L variables except L3 and L6 which are positive. Compute the relative efficiency of the store being assessed.

4.   A firm owns five residential homes for elderly people. Guests tend to fall in two categories: Those who are normally cared for by their

own family and are brought to a home for short stays and other guests who expect to stay long term in the homes. Short stay guests are charged higher prices as guest change overs involve administrative, cleaning and other costs. The homes differ in the mix of short and long term guests they serve. In order to compare the homes on their efficiency the firm collected the following data for the latest full calendar year:

Home	A	B	C	D	E
Sdays (000)	10	8	14	21	10
Ldays (000)	9	15	17	2	5
Running costs ($000)	100	85	150	90	78

Sdays and Ldays are guest days of short and long stay guests respectively. (For example one long term guest staying one week would accumulate 7 Ldays.) Running costs include staff and other incidental expenses.

a) Formulate the output maximisation envelopment model to assess the relative efficiency of home A.

b) Estimate by graphical means the levels of guest days which would render home A relatively efficient and thereby its relative efficiency rating. Explain clearly your steps.

c) The efficient peers of home A are homes B and D. What output weights would give home A its maximum relative efficiency in the value-based DEA model, where the sum of the weighted outputs is maximised?

5.   A building firm employs six representatives to go door-to-door taking orders from households for building modifications. Reviewing their performance over the first half of the latest full calendar year the firm compiled the following data:

Table 5.5.   Data on Salespersons

Representative	Hours of work (00)	Value of sales ($'000)	Households Signed Up
1	7.5	50	60
2	8	37	70
3	7	26	56
4	8.4	59	65
5	8.7	68	45
6	8.2	64	80

a) Using the data given set up a DEA model which can be used to assess the efficiency of representative 1.

b) Representative 3 was assessed by a value-based DEA model using hours worked as the sole input. It was found that she had only representative 6 as her referent efficient peer and the weight on the value of sales was approximately 0.
   (i) What is the relative efficiency of representative 3? Explain your derivation of the efficiency.
   (ii) How do you interpret the efficiency rating in (i) in terms of the output levels of representative 3?
   (iii) How is it of value to identify efficient peers in DEA? Illustrate your arguments using representative 3.

c) Discuss briefly the extent to which the factors on which data was given above are adequate for assessing the selling effectiveness of the representatives.

APPENDIX 5.1: WARWICK DEA SOFTWARE

The software is a by-product of extensive research into DEA. As the theory and reported applications of DEA advance we periodically modify the software to add to its capabilities. This Appendix summarises the capabilities of the version of the software accompanying this book. A limited User Guide can be found in Chapter 10. An upgraded version of the software is expected in the near future.

The capabilities of *Warwick-DEA Software* include:

- Assessment of units by the constant or variable returns to scale model;

- Identification of the nature of returns to scale at the efficient boundary;

- Assessment of units with restrictions to the input-output weights;

- Estimation of targets for efficiency with and without priorities over input-output improvements;

- Assessment of units when some variables are exogenously fixed and returns to scale are variable;

- Assessment of super efficiency;

- Information on the relative contributions of each peer to the targets of an inefficient unit.

Standard output includes:

- Table of efficiencies (fixed and/or spreadsheet input format);

- Table of targets (fixed and/or spreadsheet input format);

- Table of efficient peer units (fixed and/or spreadsheet input format);

- Table of virtual and raw input /output weights;

- Information on whether increasing or decreasing returns to scale hold for units efficient under variable returns to scale.

For more information on upgrades contact dea.et@btinternet.**com.**

APPENDIX 5.2

	COST	ACCOUNTS	REBATES	SUMMONS	VALUE
DEPT1	9.13	7.525	34.114	21.958	4.84
DEPT2	13.6	8.301	23.27	35.966	8.632
DEPT3	5.76	10.909	13.392	11.527	4.913
DEPT4	11.24	16.621	36.817	27.552	7.522
DEPT5	15.76	22.809	95.776	23.611	12.266
DEPT6	5.65	1.777	0.156	1.314	39.011
DEPT7	21.6	15.107	70.958	54.216	10.809
DEPT8	8.57	7.919	48.688	14.032	5.923
DEPT9	6.01	7.066	36.304	5.445	2.936
DEPT10	8.02	8.858	43.61	13.774	4.274
DEPT11	9.93	8.999	36.852	20.661	8.151
DEPT12	7.9	8.278	45.222	6.191	5.327
DEPT13	5.15	6.763	18.704	10.62	3.54
DEPT14	6.42	8.984	13.6	12.319	3.752
DEPT15	5.94	7.686	25.906	8.242	2.483
DEPT16	8.68	7.227	16.965	17.581	6.274
DEPT17	4.86	3.356	23.672	4.298	2.482
DEPT18	10.33	8.558	30.54	17.77	8.005
DEPT19	21.97	12.234	92.02	29.53	14.763
DEPT20	9.7	7.674	41.162	13.272	4.503
DEPT21	6.34	8.168	16.613	8.264	5.047
DEPT22	7.7	7.884	15.749	14.502	3.034
DEPT23	5.99	5.666	27.546	5.243	3.41
DEPT24	5.2	6.923	12.613	4.298	3.04
DEPT25	6.36	7.352	23.51	5.744	4.207
DEPT26	8.87	6.456	38.1	9.645	3.093
DEPT27	10.71	13.642	23.862	14.631	4.631
DEPT28	6.49	7.675	17.972	8.269	2.756
DEPT29	15.32	15.341	55.415	16.361	12.53
DEPT30	7	8.369	14.918	9.883	4.328
DEPT31	10.5	9.608	37.91	13.493	5.035
DEPT32	10.88	10.648	36.962	14.248	4.844
DEPT33	8.52	8.967	24.672	11.841	3.753
DEPT34	7.61	6.111	31.734	7.657	2.872
DEPT35	10.91	9.778	42.725	12.169	4.657
DEPT36	9.72	7.713	5.879	14.6	9.251
DEPT37	12.63	11.082	41.586	16.42	5.647
DEPT38	11.51	9.066	28.491	16.284	5.962

	COST	ACCOUNTS	REBATES	SUMMONS	VALUE
DEPT39	6.22	6.627	14.667	7.703	3.083
DEPT40	5.29	3.958	20.416	1.961	1.835
DEPT41	8.78	6.558	31.72	8.596	4.831
DEPT42	13.5	4.769	26.469	20.877	4.17
DEPT43	12.6	6.68	30.28	9.085	19.449
DEPT44	8.1	8.103	9.708	8.534	7.502
DEPT45	9.67	6.004	19.46	10.708	8.033
DEPT46	12.37	11.253	28.5	12.528	6.741
DEPT47	9.5	8.674	23.542	8.992	3.664
DEPT48	11.47	10.3	15.576	13.74	6.458
DEPT49	11.78	12.221	14.325	10.1	5.021
DEPT50	12.57	10.432	18.306	16.387	3.924
DEPT51	50.26	32.331	150	45.099	19.579
DEPT52	12.7	9.5	22.391	14.9	5.803
DEPT53	13.3	7.53	21.99	14.655	8.324
DEPT54	5.6	3.727	12.208	5.388	2.837
DEPT55	11.75	5.198	13.28	13.618	7.104
DEPT56	8.47	6.149	19.453	6.505	3.3
DEPT57	8.36	5.959	17.11	4.655	3.077
DEPT58	11.07	7.247	16.338	8.686	6.62
DEPT59	10.38	7.761	16.44	6.014	3.313
DEPT60	11.83	5.347	12.41	12.238	4.567
DEPT61	12.71	6.32	13.632	8.53	5.161
DEPT62	11.19	6.578	10.9	3.523	3.456

Chapter 6

DATA ENVELOPMENT ANALYSIS UNDER VARIABLE RETURNS TO SCALE

6.1 INTRODUCTION

In this chapter we relax the assumption of *constant returns to scale* (CRS) which we have maintained up to this point. It is recalled (see Chapter 3) that under CRS we assumed that if (x, y) is a feasible input-output correspondence then so is (αx, αy), where α is a non-zero positive constant. The implication of the CRS assumption can be seen readily if we consider a single-input single-output situation. If the input is x and the output is y, x being non-zero, the CRS assumption means that *average productivity*, denoted by the ratio y / x is not dependent on scale of production. For example we may be assessing a set of tax offices using their *operating expenditure* as the sole input and the *number of accounts* each tax office administers as the sole output. Then if we assume that CRS hold a Pareto-efficient tax office A which administers half the number of accounts another Pareto-efficient tax office B administers should incur half the operating expenditure of tax office B. (This assumes both tax offices face the same prices for their inputs.)

The assumption of CRS is not always appropriate in real life contexts. Even in the simple tax office example above, administering a larger number of accounts may make it possible to organise activities so as to get a rise in average productivity purely because of scale of operation, such as for example by utilising staff more fully through a steady volume of work available. CRS are impossible to sustain when scale-free variables such as indices or when arbitrary measurement scales are involved within the input-output variables. For example in assessments of school effectiveness one input used might be student attainment on entry in the form of some numerical grade while an output might be a similar grade of attainment on exit. It is then clearly not possible to argue that if two students perform 'efficiently' their grades on exit must have the same proportionality to each other as do their grades on entry. The attainment scales on entry and on exit will normally be intended to reflect different student attributes in different ways.

The most general assumption that we can make in respect of returns to scale is that they are *variable*. This permits constant but also *increasing* and *decreasing* returns to scale at different scale sizes. This chapter explains the concept of returns to scale more fully and shows how the basic DEA models introduced so far can be modified to assess performance under *Variable Returns to Scale* (VRS).

6.2   THE CONCEPT OF RETURNS TO SCALE

A starting point in coming to grips with the concept of returns to scale is that:

Returns to scale are a property of the Pareto-efficient boundary.

We address only Pareto-efficient points in reference to returns to scale because otherwise the impact of scale size on average productivity cannot be disentangled from that of gains in efficiency. Average productivity is impacted by scale size if a change in input levels leads, under efficient operation, to a non-equiproportionate change in output levels. To see this consider a single-input single-output case where some Pareto-efficient production unit has input level x, output y, and average productivity y / x. Let us change its input level marginally up, to αx, where $\alpha > 1$. Let the unit remain Pareto-efficient by changing its output level to βy. Its average productivity has now become $\beta y / \alpha x = (\beta / \alpha) \times (y / x)$. If the output of the production unit has not increased by the same proportion as its input then we will have $\alpha \neq \beta$. This will mean than the unit's new average productivity of $(\beta / \alpha) \times (y / x)$ at the new scale size of αx is not equal to its original average productivity of y / x at the scale size x. Formally:

A production correspondence is said to exhibit *increasing returns to scale* (IRS) if a radial increase in input levels (i.e. keeping input mix constant) leads under Pareto-efficiency to a more than proportionate radial increase in output levels; If the radial increase in output levels is less than proportionate we have *decreasing returns to scale* (DRS) and otherwise we have *constant returns to scale* (CRS).

Thus in the single-input single-output example above, under IRS the unit's output will rise by a larger proportion than its input and so we would have $\beta > \alpha$. This will mean $(\beta / \alpha) \times (y / x) > y/x$ and the unit will have gained in average productivity by increasing scale size from x to αx, $\alpha > 1$.

This makes clear the practical significance of identifying and exploiting returns to scale at an operating unit.

In the multi-input multi-output case the definition of IRS, CRS and DRS above in terms of the relationship between the percentage radial changes in input and output levels can be generalised as follows.

Let DMU j be Pareto-efficient and have input levels $x = \{x_{ij}, i = 1...m\}$ and output levels $y = \{y_{rj}, r = 1...s\}$. Let us scale its input levels to $\alpha x = \{\alpha x_{ij}, i = ...m\}$, where $\alpha > 0$. Let the unit be capable in principle of becoming Pareto-efficient with output levels $\beta y = \{\beta y_{rj}, r = 1...s\}$, given its input levels αx. Finally let

$\rho = \underset{\alpha \to 1}{Lim} \dfrac{\beta - 1}{\alpha - 1}$. Then

If $\rho > 1$ we have local IRS at (x, y);
If $\rho = 1$ we have local CRS at (x, y) and;   (6.1)
If $\rho < 1$ we have local DRS at (x, y).

Note that in scaling input levels by α and output levels by β above we maintain the input-output mix of DMU j constant. Thus if local IRS hold at a Pareto-efficient DMU j then when we expand (contract) its input levels by a small percentage its output levels will expand (contract) by a larger percentage, assuming the DMU remains Pareto-efficient. Under CRS the expansion (contraction) of its outputs will be by the same percentage as that of its inputs while under DRS its output levels will expand (contract) by a smaller percentage than its inputs, always assuming the DMU remains Pareto-efficient.

6.3   ASSESSING DEA EFFICIENCY UNDER VARIABLE RETURNS TO SCALE: A GRAPHICAL ILLUSTRATION

Consider a water company which has five customer services offices (SOs) whose main responsibility is the billing of clients for the water they use. Assume that the districts served by the SOs have the same mix of business and domestic clients and that all SOs provide service of similar quality. The company has collected the data in Table 6.1 on operating expenditure (OPEX) and client levels of its SOs over a certain period of time.

Table 6.1.   Data on Customer Services Offices

	OPEX ($m)	Clients (000)
SO1	2.323	201.432
SO2	0.615	119.925
SO3	0.414	49.372
SO4	0.845	70.124
SO5	2.21	249.108

Let us attempt to assess the input efficiency of SO4.

In Figure 6.1 SO1 is plotted at point SO1 and so on. If we could maintain that CRS hold, we would construct the PPS enclosed by the horizontal axis and the line OSO2. (See Figure 3.1 in Chapter 3 for a similar example.) The efficient boundary is OSO2 and its extension. On this boundary we have the maximum output to input ratio observed anywhere, namely 195 thousand clients per $1m of OPEX, found at SO2. Under input minimisation and CRS, the referent point for assessing the input efficiency of SO4 is B, a point having the same number of clients as SO4 but at a lower OPEX level. The OPEX level at B is $0.36m. This makes the input efficiency of SO4 under CRS 42.6%.

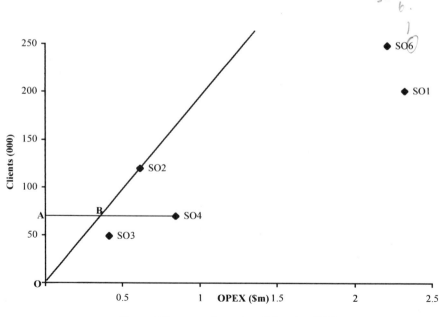

Should be 6.1

Figure 6.1.   Assessing SOs using CRS

Let us, however, assume that the SOs do not necessarily operate under CRS. Then the efficiency we obtained for SO4 above is incorrect because it relies on assuming that the ratio of 195 thousand clients per $1m of OPEX is attainable at all client levels, including the level of 70.124 thousand clients found at SO4. Yet this may well not be valid as the ratio of 195 thousand clients per $1m of OPEX represents an extrapolation from a much higher scale of operation, namely from a client level of 119.925 found at SO2. It may be that at lower client levels such as that of SO4 the efficient level of operating expenditure per client is higher because of an irreducible minimum level of set up costs of an office which are averaged across a lower number of clients. To cope with situations of this kind we need to drop the assumption of CRS and use instead an assumption of *Variable Returns to Scale* (VRS) in assessing efficiency by DEA.

We can use DEA to assess SO4 under VRS in much the same way as under CRS. We begin by constructing a PPS using all the assumptions detailed in section 3.2.1 and formally stated in Appendix 3.2, except for the assumption of CRS. The PPS under VRS for the SOs in Table 6.1 is illustrated in Figure 6.2. The piece-wise linear curve SO3SO2SO5 is constructed using the *interpolation* assumption (see Section 3.2.1.) Using the *inefficient production* assumption we construct the vertical segment down from SO3. Points along this segment are feasible but dominated by SO3. By virtue of the same assumption the horizontal segment to the right from SO5 and all points to the right of and/or below SO3SO2SO5 and its extensions are feasible.

Thus without recourse to an assumption of CRS we have constructed a PPS in Figure 6.2, consisting of SO3SO2SO5, its extensions and points to the right and or/below this piecewise linear locus. We can now measure the efficiency within this PPS in much the same way as we did in Figure 3.1 under CRS.

The input efficiency of SO4 is obtained by keeping its output constant and contracting its input as far as possible. In the context of Figure 6.2 this means that we need to identify the feasible point of lowest OPEX level along ASO4. That point is B. Expressing the input level AB as a proportion of the observed input level ASO4 of SO4 we determine that the input efficiency of SO4 under VRS is $\dfrac{AB}{ASO4}$ = 0.56, or 56%.

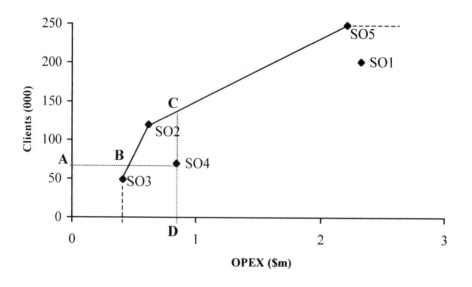

Figure 6. 2.   Assessing SOs using DEA under VRS

We had found earlier that if we assume CRS hold then SO4 should be capable of reducing its operating expenditure to 42.6% of its observed level. However, if VRS are assumed to hold then SO4 is estimated capable of reducing its operating expenditure to 56% rather than 42.6% of its observed level. It is worth clarifying how the differing assumptions over returns to scale have led to these two different estimates. As we saw earlier, the CRS-based estimate of efficient expenditure reflects the 195 thousand clients handled by SO2 per $1m expenditure. Under VRS this ratio of clients per $1m cannot be extrapolated from the 119,925 clients of SO2 to the 70,124 clients of SO4. However, the client level of SO4 is obtainable as an interpolation of the client levels of SO3 and SO2. The interpolation needed is 29.4% of SO2 and 70.6% of SO3. This interpolation gives client level of $0.294 \times 119,925 + 0.706 \times 49,372 = 70,124$. This interpolation uses input level of $0.473m (i.e. $0.294 \times 0.615m + 0.706 \times 0.414m$) which represents 56% of the observed expenditure level of SO4 of $0.845m. Thus at the heart of the difference in the VRS and CRS assessments of SO4 is that under VRS we use as benchmark output for SO4 an interpolation of output levels spanning that of SO4 rather than an extrapolation of an output level very different from that of SO4, as we did under CRS.

Note that input and output efficiency are *not* necessarily equal under VRS. For example SO4 in Figure 6.2 has output efficiency $\dfrac{DSO4}{DC} = 0.506$ some 6 percentage points less than its input efficiency. There is no contradiction here as the scale variable is the number of clients under input efficiency and operating expenditure under output efficiency. Put another way, depending on the variable used to measure scale size the DMUs compare differently on scale size and so the corresponding efficiencies can differ between input and output orientations. This means that the choice made of variable(s) for measuring scale size does impact the measure of relative efficiency of each operating unit.

6.4   ASSESSING DEA EFFICIENCY UNDER VARIABLE RETURNS TO SCALE: THE GENERIC ENVELOPMENT MODEL

It is relatively simple to modify the DEA models constructed for assessments under CRS so that they can be used to assess efficiency under VRS. Let us consider the general case in which we have N DMUs ($j = 1...N$) using m inputs to secure s outputs. Let us denote x_{ij}, and y_{rj} the level of the ith input and rth output respectively observed at DMU j. The following models can be used to assess efficiencies.

Input orientation

The *pure* technical input efficiency of DMU j_0 is h_0^*, where h_0^* is the optimal value of h_0 in

Min $h_0 - \varepsilon [\sum_{i=1}^{m} S_i^- + \sum_{r=1}^{s} S_r^+]$ [M6.1]

Subject to:

$$\sum_{j=1}^{N} \lambda_j x_{ij} = h_0 x_{ij_0} - S_i^-   i = 1...m$$

$$\sum_{j=1}^{N} \lambda_j y_{rj} = S_r^+ + y_{rj_0}   r = 1...s$$

$$\sum_{j=1}^{N} \lambda_j = 1$$

$\lambda_j \geq 0, j=1...N, S_i^-, S_r^+ \geq 0 \, \forall \, i \, and \, r$, h_0 free.

ε is a non-Archimedean infinitesimal.

Model [M6.1] was introduced in Banker et al. (1984). It differs from the generic model for assessing input efficiency under CRS (model [M4.1]) only in that it includes the so-called *convexity constraint* $\sum_{j=1}^{N} \lambda_j = 1$. This constraint is unnecessary when the DMUs operate under CRS as was shown in Chapter 3, Section 3. The convexity constraint prevents any interpolation point constructed from the observed DMUs from being scaled up or down to form a referent point for efficiency measurement since such scaling is not permissible under VRS. We can readily see the impact of the convexity constraint if we use model [M6.1] to assess SO4 within the set of offices detailed in Table 6.1.

$$\text{Min} \qquad k4 \qquad\qquad\qquad\qquad\qquad\qquad [M6.2]$$

Subject to:

$$2.323\lambda_1 + 0.615\lambda_2 + 0.414\lambda_3 + 0.845\lambda_4 + 2.21\lambda_5 = 0.845k4 - S1$$

$$201.432\,\lambda_1 + 119.925\lambda_2 + 49.372\lambda_3 + 70.124\lambda_4 + 249.108\lambda_5 = 70.124 + S2$$

$$\lambda_1 + \lambda_2 + \lambda_3 + \lambda_4 + \lambda_5 = 1$$

$$\lambda_1, \lambda_2, \lambda_3, \lambda_4, \lambda_5, S_1, S_2 \geq 0$$

At the optimal solution to this model we have $k4 = 0.559$, $\lambda_2 = 0.294$, $\lambda_3 = 0.706$. All other variables are zero. If we remove the convexity constraint $\lambda_1 + \lambda_2 + \lambda_3 + \lambda_4 + \lambda_5 = 1$ at the optimal solution to the model we will have $k4 = 0.4256$, $\lambda_2 = 0.585$. This means that in the absence of the convexity constraint we contract SO2 to 58.5% of its original scale size in input and output terms, to create a comparator relative to which we assess SO4. Under VRS we did not deem this contraction to be a legitimate action and used instead a comparator which was an interpolation of SO2 and SO3 which are respectively larger and smaller in scale size than SO4, scale size being measured on the number of clients.

The PPS under VRS is a subset of that under CRS. Otherwise the CRS and VRS DEA models work in the same way to arrive at an efficiency measure and at targets which would render a DMU Pareto-efficient. Thus, as model [M6.1] gives pre-emptive priority to the minimisation of h_0, the set of λ values which minimise h_0 to h_0^* identify a point within the VRS PPS whose input levels reflect the lowest proportion h_0^* to which the input levels of DMU j_0 can be uniformly or 'radially' contracted without detriment to its output levels.

h_0^* is termed the *pure* technical input efficiency of DMU j_0.

We refer to technical efficiencies assessed under VRS as *pure* to signal that they are 'net' of any scale effects. The impact of scale size on efficiency is measured by *scale efficiency*, which will be covered later.

We can readily deduce that at the optimal solution to [M6.1] we will have $h_0^* \leq 1$. Note that for given DMU j_0 model [M4.1] which yields its technical input efficiency can be obtained from model [M6.1] which yields its pure technical input efficiency by simply dropping the convexity constraint from the latter. This in turn can only lead to an improvement (i. e. reduction) in the optimal objective function value of model [M4.1] relative to that of model [M6.1]. Thus,

the pure technical input efficiency of a DMU is _never_ lower than its technical input efficiency.

In summary, the solution to [M6.1] yields the following information:

- DMU j_0 is Pareto-efficient if $h_0^* = 1$ and

  $S_r^{+*} = 0, r = 1....s, \ S_i^{-*} = 0, \ i = 1...m;$

- DMU j_0 has pure technical input efficiency equal to h_0^*;
- The pure technical input efficiency of DMU j_0 cannot be less than its technical input efficiency.

Target input-output levels that will render DMU j_0 Pareto efficient under VRS can be obtained from model [M6.1] in a similar manner to those detailed in expression (4.1) in Chapter 4 and derived from the input oriented DEA model [M4.1] under CRS. Models [M6.1] and [M4.1] also identify efficient peers to DMU j_0 in an identical manner. (In practice, because of the difficulty in using some value for ε in [M6.1] the model is solved by the two-phase process outlined in respect of model [M4.1] in Chapter 4.)

Example 6.1: A firm has six regional distribution centres. To assess their relative operating efficiencies it has collected the following data in respect of the last complete calendar year:

Table 6.2.   Data on Distribution Centres

Centre	Tonnes delivered (000)	Delivery points served (00)	Operating expenditure ($0,000)
1	15	10.5	38
2	14.5	12	36
3	16.5	23	42
4	18	21	45
5	21	15	37
6	11	8	17

a)   Using the above data, formulate the envelopment DEA model which can be used to assess the technical input efficiency of Centre 1.

b)   Repeat (a) in respect of the pure technical input efficiency of Centre 1.

c)   Solve the models formulated in (a) and (b) and comment on the difference in the efficiency measures you obtain.

Solution: a) It is intuitive to use the operating expenditure as an input and tonnes delivered and delivery points served as outputs. The former is to be minimised for any given levels of the two outputs.

The technical input efficiency is obtained under an assumption of CRS using the generic model in [M4.1]. The instance [M4.1] corresponding to Centre 1 is as follows:

Min $Z - \varepsilon (S1 + S2 + S3)$ [M6.3]
Subject to:

$$15\lambda_1 + 14.5\lambda_2 + 16.5\lambda_3 + 18\lambda_4 + 21\lambda_5 + 11\lambda_6 - S1 = 15$$
$$10.5 \lambda_1 + 12\lambda_2 + 23\lambda_3 + 21\lambda_4 + 15\lambda_5 + 8 \lambda_6 - S2 = 10.5$$
$$38\lambda_1 + 36\lambda_2 + 42\lambda_3 + 45\lambda_4 + 37\lambda_5 + 17\lambda_6 + S3 = 38Z$$
$$S1, S2, S3, \lambda_1, \lambda_2, \lambda_3, \lambda_4, \lambda_5, \lambda_6 \geq 0.$$

It is straightforward to see that the slack variable S3 is redundant in this single-input context as it will always be zero at the optimal solution to [M6.3].

b)   Pure technical input efficiency is obtained under VRS. It is sensible to use here an input orientation. This is consistent with treating the two output variables as exogenous and operating expenditure as the variable under managerial control. Scale size is therefore being measured here on the two output variables. As noted earlier, in the multi-output (or multi-input for that matter) context scale size is input-output mix-specific.

The generic model for assessments of pure technical input efficiency is model [M6.1]. The instance of model [M6.1] corresponding to Centre 1 is as follows:

Min $Z - \varepsilon(S1 + S2 + S3)$ [M6.4]
Subject to:

$$15\lambda_1 + 14.5\lambda_2 + 16.5\lambda_3 + 18\lambda_4 + 21\lambda_5 + 11\lambda_6 - S1 = 15$$
$$10.5\lambda_1 + 12\lambda_2 + 23\lambda_3 + 21\lambda_4 + 15\lambda_5 + 8\lambda_6 - S2 = 10.5$$
$$38\lambda_1 + 36\lambda_2 + 42\lambda_3 + 45\lambda_4 + 37\lambda_5 + 17\lambda_6 + S3 = 38Z$$
$$\lambda_1 + \lambda_2 + \lambda_3 + \lambda_4 + \lambda_5 + \lambda_6 = 1$$
$$S1, S2, S3, \lambda_1, \lambda_2, \lambda_3, \lambda_4, \lambda_5, \lambda_6 \geq 0.$$

As in model [M6.3], the slack variable S3 is redundant.

c)   The solutions to [M6.3] and [M6.4] can be obtained using suitable software such as *Lindo, ExpressMP* or an *Excel* solver. The following information is extracted from the solutions:

Model	Solution
[M6.3] (CRS)	$Z = 0.61$, $\lambda_6 = 1.36$, all other λ's zero.
[M6.4] (VRS)	$Z = 0.6578$, $\lambda_5 = 0.4$, $\lambda_6 = 0.6$, all other λ's zero.

If we assume CRS then the technical input efficiency of Centre 1 is 0.61. That is the centre could in principle reduce its operating expenditure to 61% of its observed level without detriment to its output levels. The value of 1.36 for λ_6 shows that the efficient peer on which this estimate of the efficiency of Centre 1 is based is of a smaller scale size than Centre 1 since it is being scaled up by a factor of 1.36 to provide a comparator for Centre 1. This in turn implies that Centre 1 may need to change scale of operation in order to attain the efficiency savings inherent in its efficiency measure. (We will see later how we can estimate precisely what, if any, the needed scale change is for eliminating inefficiencies due to scale of operation.)

The pure technical input efficiency of Centre 1 is yielded by model [M6.4] and it is 0.66. This means that given its scale of operation Centre 1 can lower its operating expenditure to 66% of its observed level. The values $\lambda_5 = 0.4$, and $\lambda_6 = 0.6$ show that the efficient peers on which the efficiency estimate of Centre 1 is based are Centres 5 and 6. An interpolation between these two centres, using $\lambda_5 = 0.4$, $\lambda_6 = 0.6$ yields a 'centre' of the same scale size as Centre 1 but with operating expenditure only 66% (in fact 65.78%) of that of Centre 1.

Output orientation
We can readily modify the generic model [M6.1] for assessing pure technical output efficiency. Consider the N DMUs referred to in connection with model [M6.1] and construct model [M6.5].

Max $\qquad z + \varepsilon\, [\sum_{i=1}^{m} I_i + \sum_{r=1}^{s} O_r\,]$ $\qquad$ [M6.5]

Subject to:

$$\sum_{j=1}^{N} \alpha_j x_{ij} = x_{ij_0} - I_i \qquad\qquad i = 1...m$$

$$\sum_{j=1}^{N} \alpha_j y_{rj} = O_r + z\, y_{rj_0} \qquad\qquad r = 1...s$$

$$\sum_{j=1}^{N} \alpha_j = 1$$

$\alpha_j \geq 0, j=1...N, I_i, O_r \geq 0\, \forall i\, and\, r$, z free.

ε is a non-Archimedean infinitesimal.

The optimal value $z_{j_0}^*$ of z in [M6.5] is the maximum factor by which the output levels of DMU j_0 can be radially expanded without detriment to its input levels. Thus, by definition, $\dfrac{1}{z_{j_0}^*}$ is a measure of output efficiency of DMU j_0. Since $z_{j_0}^*$ was estimated assuming VRS, $\dfrac{1}{z_{j_0}^*}$ is a measure of the *pure* technical output efficiency of DMU j_0.

In summary, using the superscript * to denote optimal values in [M6.5], it yields the following information:

- DMU j_0 is Pareto-efficient if $z_{j_0}^* = 1$ and
  $O_r^* = 0, r = 1....s, I_i^*=0, i = 1...m$;

- DMU j_0 has pure technical output efficiency of $\dfrac{1}{z_{j_0}^*}$.

- The pure technical output efficiency $\dfrac{1}{z_{j_0}^*}$ of DMU j_0 cannot be less than its technical output efficiency $\dfrac{1}{h_{j_0}^*}$ .estimated using model [M4.3].

The last point can be deduced in the same manner as for input efficiency. In practice, because of the difficulty of using some value of ε and the possibility of round off errors, model [M6.5] is solved by the two-phase process, outlined in respect of model [M4.1] in Chapter 4.

Target input-output levels that will render DMU j_0 Pareto-efficient under VRS can be obtained from model [M6.5] in a similar manner to those detailed in expression (4.2) in Chapter 4 and derived from the output oriented DEA model [M4.3] under CRS. Models [M6.5] and [M4.3] also identify efficient peers to DMU j_0 in an identical manner.

Example 6.2: Reconsider the six distribution centres of Example 6.1. Assume that the operating expenditure of each centre is relatively fixed due to contractual obligations and that the firm's aim is to maximise the tonnes delivered and the delivery points served for any given level of operating expenditure.
a)   Formulate a model to determine the technical output efficiency of Centre 1.
b)   Formulate a model to determine the pure technical output efficiency of Centre 1.
c)   Solve the models formulated and discuss any difference between the input and the output oriented efficiencies of Centre 1.

Solution: a) The generic model for estimating technical output efficiency is [M4.3]. The instance of that model corresponding to Centre 1 of Example 6.1 is as follows:

Max $h + \varepsilon(S1 + S2 + S3)$ [M6.6]
Subject to:
$15\alpha_1 + 14.5\alpha_2 + 16.5\alpha_3 + 18\alpha_4 + 21\alpha_5 + 11\alpha_6 - S1 = 15h$
$10.5\alpha_1 + 12\alpha_2 + 23\alpha_3 + 21\alpha_4 + 15\alpha_5 + 8\,\alpha_6 - S2 = 10.5h$
$38\alpha_1 + 36\alpha_2 + 42\alpha_3 + 45\alpha_4 + 37\alpha_5 + 17\alpha_6 + S3 = 38$
$S1, S2, S3, \alpha_1, \alpha_2, \alpha_3, \alpha_4, \alpha_5, \alpha_6 \geq 0.$

b)   The generic model for determining the pure technical output efficiency of Centre 1 is model [M6.5]. The instance of that model corresponding to Centre 1 of Example 6.1 is as follows:

Max $h + \varepsilon(S1 + S2 + S3)$ [M6.7]
Subject to:

$$15\alpha_1 + 14.5\alpha_2 + 165\alpha_3 + 18\alpha_4 + 21\alpha_5 + 11\alpha_6 - S1 = 15h$$
$$10.5\alpha_1 + 12\alpha_2 + 23\alpha_3 + 21\alpha_4 + 15\alpha_5 + 8\alpha_6 - S2 = 10.5h$$
$$38\alpha_1 + 36\alpha_2 + 42\alpha_3 + 45\alpha_4 + 37\alpha_5 + 17\alpha_6 + S3 = 38$$
$$\alpha_1 + \alpha_2 + \alpha_3 + \alpha_4 + \alpha_5 + \alpha_6 = 1$$
$$S1, S2, S3, \alpha_1, \alpha_2, \alpha_3, \alpha_4, \alpha_5, \alpha_6 \geq 0.$$

c)   Solving model [M6.6] yields $h^* = 1.639$. This makes the technical output efficiency of Centre 1, $\dfrac{1}{1.639} = 0.61$. Solving model [M6.7] yields $h^* = 1.4$. This makes the pure technical output efficiency of Centre 1, $\dfrac{1}{1.4} = 0.714$.

The technical efficiency measure is the same whether measured in the input or the output orientation. In the case of Centre 1 in both cases it is 0.61. This is as expected. In contrast, the pure technical input efficiency of Centre 1 is 0.6578 (see the solution to model [M6.4]) while its pure technical output efficiency is 0.714. The centre has different scale size in the input and the output orientation. In the input orientation an interpolation of Centres 5 and 6 is needed to obtain a centre of the same output levels and mix as Centre 1, creating thus a comparator of the same scale size as Centre 1. In contrast, in the output orientation, Centre 5 which has almost the same operating expenditure as Centre 1 constitutes the comparator for Centre 1. These differing views on the scale size of Centre 1 between the input and output orientations explain the difference in its efficiency measures in the two orientations.

6.5   VALUE-BASED DEA MODELS UNDER VRS

As in the case under CRS (see Chapter 4) so too under VRS, we can use value-based DEA models to assess efficiency. Such models are dual, in linear programming terms, to the generic envelopment DEA models [M6.1] and [M6.5] described above. (See Appendix 3.1 for a brief introduction to duality in Linear Programming.)

Consider first the envelopment model [M6.1] used to assess pure technical input efficiency. The corresponding value-based DEA model for assessing the pure technical input efficiency of DMU j_0 is as follows.

$$\text{Max} \qquad p_0 = \sum_{r=1}^{s} u_r y_{rj_0} + \omega \qquad\qquad [\text{M6.8}]$$

subject to
$$\sum_{i=1}^{m} v_i x_{ij_0} = 1$$

$$\sum_{r=1}^{s} u_r y_{rj} - \sum_{i=1}^{m} v_i x_{ij} + \omega \le 0 \qquad j = 1...j_0...N$$

$$u_r \ge \varepsilon \qquad\qquad r = 1...s$$

$$v_i \ge \varepsilon, \qquad\qquad i = 1...m$$

$$\omega \text{ free.}$$

where ω, u_r and v_i are variables. Notation in [M6.8] is otherwise as in [M6.1]. Let the superscript * denote the optimal value of the corresponding variable in [M6.8]. The pure technical input efficiency of DMU j_0 as yielded by model [M6.8] is p_0^*.

DMU j_0 is Pareto-efficient if and only if $p_0^* = 1$.

The variables u_r and v_i in [M6.8] reflect respectively imputed values on the rth output and ith input. The values of u_r and v_i, when not infinitesimal, can be used to arrive at marginal rates of substitution between inputs or outputs, or marginal rates of transformation between inputs and outputs in a manner analogous to that outlined with reference to the optimal u_r and v_i of model [M4.5] in Chapter 4. Model [M6.8] differs from the corresponding value-based DEA model in [M4.5] only in that [M6.8] involves the additional variable ω which is dual (in linear programming terms) to the convexity constraint in [M6.1]. The value of this variable, as we will see later in this chapter, reflects the impact of scale size on the productivity of a DMU. If ω is zero at an optimal solution to model [M6.8] the model collapses to [M4.5], the value-based DEA model under CRS. In such a case DMU j_0 lies on or is projected at a point on the Pareto-efficient boundary where locally CRS hold. If the optimal value of ω is not zero it may be positive or negative, its sign depending, as we will see below, on the type of returns to scale holding locally where DMU j_0 lies or is projected on the efficient boundary.

Consider now the envelopment model [M6.5] used to assess pure technical output efficiency. The corresponding value-based DEA model for assessing the pure technical output efficiency of DMU j_0 is as follows.

Min $\qquad h_{j_0} = \sum_{i=1}^{m} \delta_i x_{ij_0} + w$ $\qquad\qquad$ [M6.9]

Subject to:

$$\sum_{r=1}^{s} \gamma_r y_{rj_0} = 1$$

$$\sum_{r=1}^{s} \gamma_r y_{rj} - \sum_{i=1}^{m} \delta_i x_{ij} - w \leq 0 \qquad\qquad j = 1...j_0...N$$

$$\gamma_r \geq \varepsilon \qquad\qquad\qquad\qquad\qquad r = 1...s$$

$$\delta_i \geq \varepsilon, \qquad\qquad\qquad\qquad\qquad i = 1...m.$$

w free.

Where w, γ_r and δ_i are variables. Notation in [M6.9] is otherwise as in [M6.5]. We shall use the superscript * to denote the optimal value of the corresponding variable in [M6.9]. The pure technical output efficiency of DMU j_0 is $1 / h_{j_0}^*$.

DMU j_0 is Pareto-efficient if and only if $h_{j_0}^*$ = 1.

Example 6.3: Consider the 9 DMUs in Table 6.3 and assume that they operate an input to output transformation process characterised by VRS.

Table 6.3. $\qquad$ Sample Input-Output Data$^\$$

	DMU 1	DMU 2	DMU 3	DMU 4	DMU 5	DMU 6	DMU 7	DMU 8	DMU 9
Input 1	10	15	35	30	15	60	10	18.75	30
Input 2	15	10	20	25	15	60	10	18.75	27.5
Output	10	10	20	20	10	30	7	15	20

$ Numerical data from Appa and Yue (1999)

Use a value-based DEA model to ascertain the technical input efficiency of DMU 6.

Solution: The value-based generic DEA model for assessing pure technical input efficiency is [M6.8]. Its instance in respect of DMU 6 can be written as follows:

Max $30u + w1 - w2$ [M6.10]
Subject to:

$$60v1 + 60v2 = 100$$
$$10u - 10v1 - 15v2 + w1 - w2 \leq 0$$
$$10u - 15v1 - 10v2 + w1 - w2 \leq 0$$
$$20u - 35v1 - 20v2 + w1 - w2 \leq 0$$
$$20u - 30v1 - 25v2 + w1 - w2 \leq 0$$
$$10u - 15v1 - 15v2 + w1 - w2 \leq 0$$
$$30u - 60v1 - 60v2 + w1 - w2 \leq 0$$
$$7u - 10v1 - 10v2 + w1 - w2 \leq 0$$
$$15u - 18.75v1 - 18.75v2 + w1 - w2 \leq 0$$
$$20u - 30v1 - 27.5v2 + w1 - w2 \leq 0$$
All variables ≥ 0.

(Note that within model [M6.10] we have normalised total virtual input to 100 to reduce the impact of round off errors and we have used the difference of two non-negative variables (w1 - w2) to model the free variable ω of model [M6.8].)

Part of the optimal solution to model [M6.10], obtained by Lindo is as follows.

Objective function value: 100.0000

VARIABLE	VALUE
u	5.0
w1	0.0
w2	50.0
v1	1.6667
v2	0.0

Thus the technical input efficiency of DMU 6 is 100 (%). The DMU is Pareto-efficient. The ω value is -50 which as we will see in the next section can be used to identify the type of returns to scale holding locally at DMU 6.

6.6 SCALE EFFICIENCY, RETURNS TO SCALE AND MOST PRODUCTIVE SCALE SIZE

Given that under VRS scale size affects the productivity of a DMU some important questions arise in respect of the scale size at which a DMU operates. For example is there a scale size that would be optimal in some sense for the DMU? Is there a measure of the impact of scale size on the DMU's productivity? Questions of this type are captured in the concepts of *'scale efficiency'*, *'returns to scale'* and *'most productive scale size'*. This section covers the use of DEA models in the context of these concepts, drawing largely on the work of Banker et al. (1984), Banker (1984) and Banker and Thrall (1992).

6.6.1 Scale Efficiency

Scale efficiency measures the impact of scale size on the productivity of a DMU. It is defined as follows:

Scale input efficiency of DMU j_0 =

$$\frac{Technical\,input\,efficiency\,of\,DMU\ j_0}{Pure\,technical\,input\,efficiency\,of\,DMU\ j_0} \qquad (6.2).$$

Scale output efficiency is defined in an analogous manner.

As we saw earlier, the technical efficiency of a DMU can never exceed its pure technical efficiency in either orientation. Thus from the definition of scale efficiency in (6.2) we conclude that we always have

Scale efficiency ≤ 1   (6.3).

By definition, scale efficiency measures the divergence between the efficiency rating of a DMU under CRS and VRS respectively. Unlike the CRS efficiency rating, the VRS rating is obtained when we control for the scale size of the DMU. This is the only difference in how the two measures of efficiency are obtained and so the divergence of the measures captures the impact of scale size on the productivity of the DMU concerned.

The impact of scale size on productivity and the measurement of this impact by scale efficiency can be illustrated graphically. Reconsider the five customer service offices (SOs) of Table 6.1. They are plotted in Figure 6.3. When CRS are maintained the locus of efficient DMUs is OSO2 and the PPS is to the right and below that line. In contrast, when VRS are maintained the locus of efficient DMUs is SO3SO2SO5 and the PPS is to the right and below that piece-wise linear boundary.

Consider now SO4 plotted at point SO4 in Figure 6.3. Its technical input efficiency (under CRS) is $\dfrac{AC}{ASO4}$ while its pure technical input efficiency (under VRS) is $\dfrac{AB}{ASO4}$. Thus its scale efficiency is $\dfrac{AC}{ASO4} \div \dfrac{AB}{ASO4} = \dfrac{AC}{AB}$. This ratio is a measure of the distance between the points B and C in Figure 6.3. This distance is the expenditure saving SO4 could have made if it had been operating at point C which would mean that it would have served the same number of clients per $1m as SO2. This potential saving is not realised by SO4 purely because of the scale (in terms of clients) it operates at since even if it had been efficient at its current scale size it would have only

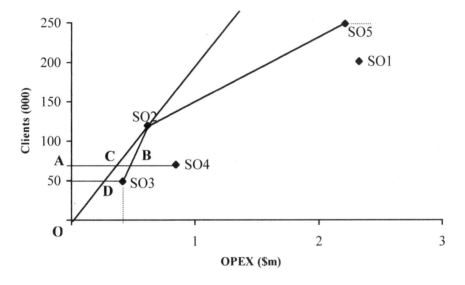

Figure 6 3.   Interpreting Scale Efficiency

operated at point B.

The scale input efficiency of SO3 is $\dfrac{\text{Technical input efficiency of SO3}}{\text{Pure technical input efficiency of SO3}} = \dfrac{0.61}{1} = 0.61$. In contrast, the input scale efficiency of SO4 is $\dfrac{0.426}{0.56} = 0.76$. (The efficiencies used here have been computed using software.) The larger input scale efficiency of SO4 suggests that scale size in itself impacts more favourably the productivity of SO4 than of SO3. This indeed can be seen in Figure 6.3

where the waste of resource attributable to scale size is DSO3 in the case of SO3 and CB in the case of SO4 and we have CB < DSO3.

If the technical and pure technical efficiency of a DMU are equal then scale efficiency is 1 and whether or not we control for its scale size we reach the same view on the DMU's technical efficiency. We can identify no adverse impact of scale size on its productivity. If however, the DMU has lower CRS compared to VRS efficiency then its scale efficiency will be below 1. The lower CRS compared to VRS efficiency rating suggests the Pareto-efficient comparator to the DMU is more productive in the former case and less productive when we control for scale size which means that scale of operation does impact the productivity of the DMU. Thus:

The larger the divergence between VRS and CRS efficiency ratings the lower the value of scale efficiency and the more adverse the impact of scale size on productivity.

6.6.2 Identifying Returns to Scale by Means of DEA Models

We saw earlier in this chapter that returns to scale are to do with how average productivity is affected by scale size and that they are a property of the Pareto-efficient boundary. We also saw that if increasing returns to scale (IRS) hold at a production point then raising its input levels by a small percentage will lead to an expansion of its output levels by an even larger percentage, assuming the unit remains Pareto-efficient. The percentage rise in output levels will be lower than that of the input levels if decreasing returns to scale (DRS) hold while inputs and outputs will change by the same percentage if CRS hold.

In the single-input single-output case we can easily see the type of returns to scale holding at each part of the efficient boundary. Figure 6.4 depicts an illustrative set of DMUs where the PPS under VRS is ABCD, its vertical and horizontal extensions and the space to the right and below. Take the Pareto-efficient segment AB. The expression linking *Output* to *Input* along the segment has the form

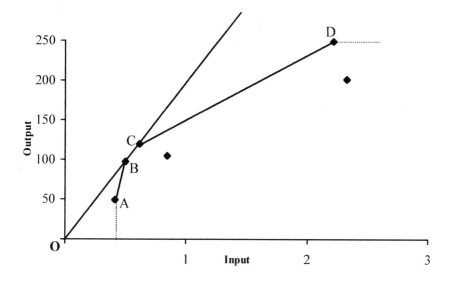

Figure 6.4.   Illustrating IRS, CRS and DRS

$$Output = a\ Input + b \tag{6.4}$$

where a is the slope of AB and b is the intercept on the *Output* axis. **Note that b is negative.** If we raise the level of *Input* at A by say 10% then to retain Pareto efficiency and remain on AB the corresponding *Output* level would rise to

$$Output(1.1A) = a \times 1.1 \times Input_A + b \tag{6.5}$$

where $Input_A$ is the level of Input at A and $Output(1.1A)$ is the level of *Output* when *Input* is 10% above that at A. Note, however, that

$$\frac{Output(1.1A)}{Output_A} = \frac{a1.1\,Input_A + b}{a\ Input_A + b} = \frac{a\ Input_A + b}{a\ Input_A + b} +$$

$$0.1\frac{a\ Input_A + b}{a\ Input_A + b} - 0.1\frac{b}{a\ Input_A + b} = 1.1 - 0.1\frac{b}{a\ Input_A + b} \tag{6.6}.$$

The last term in (6.6) is positive because b is negative and $(a\ Input_A + b)$ is positive. This means that the 10% rise in input at A has led to a more than 10% rise in the level of *Output*. So

along the segment AB we have IRS.

If we repeat the foregoing procedure at point B in respect of the segment BC, the last term in the equivalent to (6.6) expression will be zero. This is because the segment BC in Figure 6.4 goes through the origin and so has intercept b = 0. Thus a 10% rise in input B would have led to a 10% rise in output along BC. So

along the segment BC we have CRS.

Finally, if we follow at C the same procedure in respect of segment CD as we did in respect of segment AB above the last term in the equivalent expression to (6.6) would be negative as the intercept of CD, (b in 6.4) is positive. This means a 10% rise in input at C would have led along CD to a less than 10% rise in output. So

along the segment CD we have DRS.

This type of information is very important for managerial decision making. Obviously it makes sense for a DMU operating at a point where IRS hold to increase its scale size, if this is under its control in any way, as its additional input requirements will be more than compensated for by a rise in output levels. Similarly, a DMU operating at a point where decreasing returns to scale hold should decrease its scale size. The ideal scale size to operate at is where CRS hold. This scale size which it is recalled is input-output mix specific is what Banker (1984) termed *'most productive scale size'*, (MPSS). We shall return to the identification of this scale size in the general case after we first look at how we can identify the type of returns to scale holding in the general case.

The issue of identifying the type of returns to scale holding at the locality of a DMU was first addressed by Banker (1984). He used the sum of λ values at the optimal solution to the generic envelopment model under CRS, model [M4.1], to identify the nature of the returns to scale at the locality of a DMU. However, there could be multiple optimal solutions to [M4.1] and this point was taken on board later by Banker and Thrall (1992) on whose work we now draw to state a set of conditions for identifying the nature of returns to scale holding at DMU j_0. We can use either envelopment or value-based DEA models to identify the nature of returns to scale holding locally at DMU j_0. The conditions are stated here. (The relevant proofs are beyond the scope of this book but can be found by the interested reader in the original work in Banker and Thrall (1992).)

Using Envelopment DEA Models

Input orientation

Consider a set of DMUs (j = 1...N) operating under VRS and let DMU j_0 be Pareto-efficient. Solve model [M4.1] in respect of DMU j_0 and let the superscript * to a variable denote its optimal value in [M4.1].

(6.7)

- *If $\sum \lambda_j^* > 1$ for <u>all</u> optimal solutions to [M4.1] then DRS hold locally at DMU j_0;*
- *If $\sum \lambda_j^* = 1$ for <u>at least one</u> optimal solution to [M4.1] then CRS hold locally at DMU j_0;*
- *If $\sum \lambda_j^* < 1$ for <u>all</u> optimal solutions to [M4.1] then IRS hold locally at DMU j_0.*

Output orientation

The nature of returns to scale is identified using the conditions in (6.7) except that we solve the output oriented CRS model [M4.3] instead of [M4.1] and use in the context of (6.7) $\sum \alpha_j^*$ instead of $\sum \lambda_j^*$.

Using value-based DEA models

Input orientation

Consider a set of DMUs (j = 1...N) operating under VRS. Solve model [M6.8] in respect of DMU j_0 and let DMU j_0 be Pareto-efficient:

(6.8)

- *If ω takes negative values in <u>all</u> optimal solutions to model [M6.8] then locally at DMU j_0 decreasing returns to scale hold;*
- *If ω takes a zero value in <u>some</u> optimal solution to [M6.8] then locally where DMU j_0 lies or is projected on the efficient boundary CRS hold;*
- *If ω takes positive values in <u>all</u> optimal solutions to [M6.8] then locally at DMU j_0 increasing returns to scale hold.*

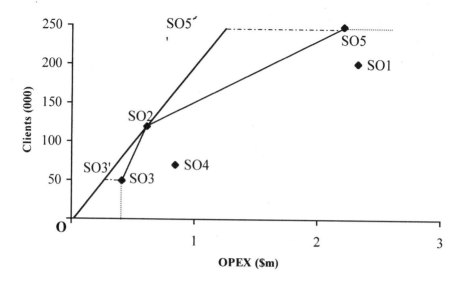

Figure 6.5.   Illustrating the Conditions in (6.7)

Output orientation

Consider a set of DMUs (j = 1...N) operating under VRS. Solve model [M6.9] in respect of DMU j_0 and let DMU j_0 be Pareto-efficient:

$$(6.9)$$

- *If w takes negative values in __all__ optimal solutions to model [M6.9] then locally at DMU j_0 increasing returns to scale hold;*
- *If w takes a zero value in __some__ optimal solution to [M6.9] then locally where DMU j_0 lies or is projected on the efficient boundary CRS hold.*
- *If w takes positive values in __all__ optimal solutions to [M6.9] then locally at DMU $j0$ decreasing returns to scale hold.*

We can readily see how the conditions in (6.7) work in the case of Figure 6.3, reproduced for ease of reference here as Figure 6.5. If we solve [M4.1] at SO3 the reference point will be SO3'. This point represents a contraction of SO2 and so [M4.1] will yield a single positive λ, that corresponding to SO2. That λ will be less than 1 to reflect a contraction of SO2 to the point SO3'. Under the conditions in (6.7), the λ being less than 1 would lead us to conclude that SO3 operates under increasing returns to scale. This is indeed the case as we can conclude from the fact that the segment SO3SO2 has a negative intercept. (See point A in Figure 6.4 which is analogous to point SO3 in Figure 6.5.) Similarly we can easily see that if we solve [M4.1] in respect of SO5 only the λ corresponding to SO2 will be positive and it will be above 1, scaling up SO2 to the point SO5'. The λ in excess of 1 would according to (6.7) lead us to conclude that SO5 operates at a part of the VRS boundary where decreasing returns to scale hold. This we know to be true from the fact that SO2SO5 has a positive intercept. (See the discussion relating to Figure 6.4, segment CD above). Finally if we solve [M4.1] with respect to SO2 the λ relating to that SO2 will be 1. This again signals that SO2 is on the part of the VRS boundary where CRS hold. (In Figure 6.5 CRS hold only at point SO2 whereas in Figure 6.4 all points on the segment BC operate under CRS.

6.6.3 Most Productive Scale Size

It is intuitively obvious that a Pareto-efficient DMU which operates under IRS would gain in average productivity if it increases marginally its scale size. This is because if the unit remains Pareto-efficient, proportional increases in its input levels will be followed by rises in its output levels by a larger percentage than was the case for the input levels. Following a similar argument we can deduce that a Pareto-efficient DMU operating under DRS would become more productive if it were to decrease its scale. Only for DMUs operating under CRS will productivity be unaffected by marginal changes in scale size. Thus any Pareto-efficient DMU not operating under local CRS will gain in productivity if it changes its scale size up to the point where returns to scale become constant.

The foregoing intuitive argument leads to the conclusion that the 'optimal' scale size to operate at is where local CRS hold. Such scale size is known as '*most productive scale sizes' (MPSS)*. Formally (see Banker and Thrall (1992) (Proposition 1)) one way to define MPSS is as follows:

A production possibility is MPSS if and only if it has pure technical and scale efficiency of 1.

One obvious corollary of this definition is that

if a DMU is Pareto-efficient under CRS it will have MPSS.

Note that the converse is not true. That is to say a DMU could be MPSS but not necessarily Pareto-efficient. A second fairly obvious corollary of the definition of MPSS above is that

the average productivity of an input-output mix is maximum at the MPSS for that input-output mix.

This corollary is fairly easy to see in the single-input single-output case. For example in Figure 6.5 we have MPSS at SO2, the only point where CRS hold and where average productivity (clients per $1m of OPEX) is maximum while in Figure 6.4, there is a range of MPSS, those on the segment BC where again average productivity is maximum. In the general multi-input multi-output case the corollary becomes easier to see if we note the following necessary condition for MPSS (Cooper et al. (2000) (Model 5.22)):

Consider N DMUs (j = 1...N) operating under VRS and using m inputs to secure s outputs. Let us denote x_{ij} and y_{rj} the level of the ith input and rth output respectively observed at DMU j. Solve in respect of DMU j_0 the model:

Max   b / a   [M6.11]
Subject to:

$$\sum_{j=1}^{N} \lambda_j x_{ij} \leq a x_{ij_0} \qquad i = 1...m$$

$$\sum_{j=1}^{N} \lambda_j x_{ij} \geq b y_{rj_0} \qquad r = 1...s$$

$$\sum_{j=1}^{N} \lambda_j = 1$$

$$\lambda_j \ (j = 1...N), a, b \geq 0$$

A necessary condition for DMU j_0 to be MPSS is that the optimal objective function value b* / a* in [M511] be 1.

[M6.11]

We can view b / a as defined in [M6.11] as 'average productivity' of an input-output mix in the multi-input multi-output case since the ratio b / a is of 'output' to 'input' keeping the input-output mix constant. This ratio is maximum at an MPSS.

In the general case we can identify the Pareto-efficient MPSS for the input-output mix of a DMU as follows:

Input orientation

Solve model [M4.1] (the generic CRS envelopment model ib Chapter 4) in respect of DMU j_0 and let the superscript * to a variable denote its optimal value in [M4.1]. Each set of λ values which is consistent with $k_0 = k_0^*$ in [M4.1] corresponds to a set of Pareto-efficient MPSS input-output levels (X^{MPSS}, Y^{MPSS}) where $X^{MPSS} = (\ x_{ij_0}^{MPSS}\ , i = 1...m)$, $Y^{MPSS} = (\ y_{rj_0}^{MPSS}\ , r = 1...s)$ and

$$x_{ij_0}^{MPSS} = \frac{\sum_{j=1}^{N} \lambda_j^* x_{ij}}{\sum_{j=1}^{N} \lambda_j^*} \qquad i = 1...m$$

$$y_{rj_0}^{MPSS} = \frac{\sum_{j=1}^{N} \lambda_j^* y_{rj}}{\sum_{j=1}^{N} \lambda_j^*} \qquad r = 1...s. \qquad (6.10)$$

Output orientation

Pareto-efficient MPSS input-output levels for DMU j_0 above can be computed using the conditions stated in (6.10) except that we solve the output oriented envelopment CRS model [M4.3] instead of [M4.1], using in the context of (6.10) $h_{j_0} = h_{j_0}^*$ instead of $k_0 = k_0^*$ and $\sum \alpha_j^*$ instead of $\sum \lambda_j^*$.

As it is clear from the foregoing, when there are multiple optimal solutions to [M4.1] or [M4.3] (6.10) will yield multiple sets of MPSS input-output correspondences. Appa and Yue (1999) have developed a model for estimating what they call '*best returns to scale*' MPSS input-output levels. These correspond to the smallest or alternatively to the largest scale size that is MPSS (e.g. points B or C in Figure 6.4). (See also Zhu (2000) on this point.)

Example 6.4: Consider the 9 DMUs in Table 6.3 and assume that they operate an input to output transformation process characterised by VRS.

a) Determine whether DMUs 6 and 9 are Pareto-efficient.

b) Identify the type of returns to scale holding locally at any Pareto-efficient DMU(s) you identify in (a).

c) Identify a MPSS set of input-output levels for DMU 6 in the input orientation.

Solution: a) We have already identified in Example 6.3 that DMU 6 is Pareto-efficient under VRS. Turning to DMU 9, the generic model for assessing pure technical input efficiency is model [M6.1]. The instance of model [M6.1] corresponding to DMU 9 is

MIN $Z - 0.001S1 - 0.001S2 - 0.001SO$ [M6.12]

Subject to:

INPUT 1: $10L1 + 15L2 + 35L3 + 30L4 + 15\,L5 + 60L6 + 10\,L7 + 18.75L8 + 30L9 + S1 - 30Z = 0$

INPUT 2: $15L1 + 10L2 + 20L3 + 25L4 + 15L5 + 60L6 + 10L7 + 18.75L8 + 27.5L9 + S2 - 27.5Z = 0$

OUTPUT: $10L1 + 10L2 + 20L3 + 20L4 + 10L5 + 30L6 + 7L7 + 15L8 + 20L9 - SO = 20$

CONVEXITY: $L1 + L2 + L3 + L4 + L5 + L6 + L7 + L8 + L9 = 1$

             Z free, all other variables non-negative.

This model can be solved using software such as Lindo. At the optimal solution we have $Z = 1$, $S1 = SO = 0$ and $S2 = 2.5$. Thus DMU 9 is not Pareto-efficient though its radial efficiency is 1. There is slack in input 2.

b) We have only identified DMU 6 as Pareto-efficient. To identify the nature of returns to scale locally we can use either the input or the output oriented envelopment DEA model under CRS. For example we can solve the instance of model [M4.1] corresponding to DMU 6. The resulting model is [M6.13].

MIN $Z - 0.001S1 - 0.001S2 - 0.001SO$ [M6.13]

Subject to:

INPUT 1: $10L1 + 15L2 + 35L3 + 30L4 + 15\,L5 + 60L6 + 10L7 + 18.75L8 + 30L9 + S1 - 60Z = 0$

INPUT 2: $15L1 + 10L2 + 20L3 + 25L4 + 15L5 + 60L6 + 10L7 + 18.75L8 + 27.5L9 + S2 - 60Z = 0$

OUTPUT: $10L1 + 10L2 + 20L3 + 20L4 + 10L5 + 30L6 + 7L7 + 15L8 + 20L9 - SO = 30$

             Z free, all other variables non-negative.

The positive values obtained are: $Z = 0.625$ and $L1 = L2 = 1.5$. All other variables are zero. In order to use the conditions in (6.7) to identify the type of returns to scale holding at DMU 6 we need to know if there exists any optimal solution to [M6.13] in which the L's sum to 1. We can reformulate

[M6.13] to minimise the sum of L's, setting in the constraints of the reformulated model the optimal value of Z obtained earlier, which is Z = 0.625. If the sum of L's turns out to be 1 or less then we will have established that there exists an optimal solution to [M6.13] in which the L's sum to 1. In the case of [M6.13] the minimum sum of L's compatible with Z = 0.625 is 2.

Since at the optimal solution to [M6.13] the sum of positive L's is 2 > 1, and there exists no optimal solution where the L's sum to 1, then using the conditions in (6.7), we conclude that in the input orientation DMU 6 is operating under decreasing returns to scale.

c)   To identify a set of MPSS levels for DMU 6 we use expression (6.10). One optimal sum of L's in [M6.13] was 3 and the positive L's were L1 = L2 = 1.5. This leads, to the following MPSS, based on the expression in (6.10):

Input 1: MPSS Level = (1.5x10 + 1.5x15) / 3 = 12.5
Input 2: MPSS Level = (1.5x15 + 1.5x10) / 3 = 12.5
Output: MPSS Level = (1.5x10 + 1.5x10) / 3 = 10.0

We have alternative MPSS levels for DMU 6 since we had alternative optimal solutions to [M6.13].

Example 6.5: Reconsider the 9 DMUs in Table 6.3 and maintain the assumption that they operate an input to output transformation process characterised by VRS. Use a <u>value-based</u> DEA model to ascertain the nature of returns to scale holding locally at DMU 6.

Solution: The generic value-based DEA model for assessing pure technical input efficiency is [M6.8]. Its instance in respect of DMU 6 in Table 6.3 can be written as in [M6.14]. (Note that in deriving [M6.14] from model [M6.8] we have normalised total virtual input to 100 to reduce the impact of round off errors and we have used the difference of two non-negative variables (w1-w2) to model the free variable ω.)

Max $30u + w1-w2$ [M6.14]
Subject to:

$$60v1 + 60v2 = 100$$
$$10u-10v1-15v2 + w1-w2 \leq 0$$
$$10u-15v1-10v2 + w1-w2 \leq 0$$
$$20u-35v1-20v2 + w1-w2 \leq 0$$
$$20u-30v1-25v2 + w1-w2 \leq 0$$
$$10u-15v1-15v2 + w1-w2 \leq 0$$
$$30u-60v1-60v2 + w1-w2 \leq 0$$
$$7u-10v1-10v2 + w1-w2 \leq 0$$
$$15u-18.75v1-18.75v2 + w1-w2 \leq 0$$
$$20u-30v1-27.5v2 + w1-w2 \leq 0$$

All variables ≥ 0.

Part of the optimal solution to model [M6.14], obtained by Lindo is as follows.

Objective function value: 100.0000

VARIABLE	VALUE
u	5.0
w1	0.0
w2	50.0
v1	1.6667
v2	0.0

We need to ascertain whether there are alternative optimal solutions to the above and, if so, whether at any one of these optimal solutions w1-w2 becomes zero. As we have found w1-w2 to be negative we can test whether at some optimal solution w1-w2 becomes zero by solving model [M6.15].

At the optimal solution to this model we have w1 = 0, w2 = 50, in fact the same solution as to model [M6.14] above. Thus there exists no optimal solution to model [M6.14] where w1-w2 is 0. Since at all optimal solutions to [M6.14] w1-w2 takes negative values we conclude from the conditions in (6.8) that DMU 6 lies on a part of the efficient boundary where decreasing returns to scale hold.

Max   w1 - w2   [M6.15]

Subject to:

$$30u + w1 - w2 \geq 100$$
$$60v1 + 60v2 = 100$$
$$10u - 10v1 - 15v2 + w1 - w2 \leq 0$$
$$10u - 15v1 - 10v2 + w1 - w2 \leq 0$$
$$20u - 35v1 - 20v2 + w1 - w2 \leq 0$$
$$20u - 30v1 - 25v2 + w1 - w2 \leq 0$$
$$10u - 15v1 - 15v2 + w1 - w2 \leq 0$$
$$30u - 60v1 - 60v2 + w1 - w2 \leq 0$$
$$7u - 10v1 - 10v2 + w1 - w2 \leq 0$$
$$15u - 18.75v1 - 18.75v2 + w1 - w2 \leq 0$$
$$20u - 30v1 - 27.5v2 + w1 - w2 \leq 0$$

All variables ≥ 0.

6.7 PRACTICAL USE OF DEA MODELS UNDER VRS

Chapter 5 detailed how DEA models under CRS are most commonly put to use. DEA models under VRS can be put to use in much the same way. Moreover, now that DMUs operate under VRS, the DEA models can additionally be used to identify the type of returns to scale holding locally at the efficient boundary where a DMU lies or is projected, and to identify a most productive scale size for the DMU. We briefly outline here how the uses of DEA models detailed in Chapter 5 can be implemented and indeed augmented where DEA models under VRS are concerned.

6.7.1 Envelopment models

The envelopment models, [M6.1] and [M6.5] are most appropriate for:

– Ascertaining the pure technical input or output efficiency of the DMU.

– Getting a view on the scope for the DMU to be a role-model for other DMUs.
This is appropriate where the DMU is Pareto-efficient. The proportion of inefficient DMUs which have the DMU concerned as an efficient peer and its contribution to the aggregate target levels of inefficient DMUs jointly convey some measure of the scope for the efficient DMU to be a role model for less efficient DMUs. Peer identification and computation

of contribution to targets is done in the same manner under VRS as detailed in Chapter 5 for DEA models under CRS.

– Identifying efficient DMUs whose operating practices an inefficient DMU may attempt to emulate to improve its performance. Efficient peers to an inefficient DMU are identified in the same manner in DEA models under VRS as under CRS.

– Estimating target input-output levels which an inefficient DMU should in principle be capable of attaining under efficient operation.
Target input-out put levels which would render DMU j_0 Pareto-efficient under VRS are identified in the input orientation by applying expression (4.1) of Chapter 4 to the solution of model [M6.1], using h_0^* in place of k_0^*. Target input-output levels which would render DMU j_0 Pareto-efficient under VRS in the output orientation are identified by applying expression (4.2) of Chapter 4 to model [M6.5] using z^* instead of $h_{j_0}^*$ within (4.2).

– Ascertaining the nature of returns to scale holding locally at the efficient boundary where a DMU lies or is projected. The conditions used for this were detailed in (6.7).

– Identifying the most productive scale size for a DMU. The input-output levels which would give a DMU its most productive scale size were defined in (6.10).

6.7.2 Value based models

The value-based DEA models, [M6.8] and [M6.9] are most appropriate for:
– Ascertaining the pure technical input or output efficiency of each DMU;

– Getting a view on the robustness of the pure technical efficiency of a DMU.
Such information is conveyed by the virtual input and output levels of the DMU being assessed in the same manner as under CRS. However, there is now an additional 'virtual' element in the form of ω in [M6.8] or w in [M6.9]. Unless the DMU concerned operates, or is projected to a point on the efficient boundary where CRS hold, ω in [M6.8] or w in [M6.9] would be non-zero. This means that in part the pure technical efficiency

rating of the DMU concerned would be 'explained' by the scale at which the DMU operates. So one way to view for example ω in the context of model [M6.8] is as a measure of the degree to which the pure technical input efficiency of DMU j_0 reflects its scale size rather than output volumes attained relative to input levels. Similar comments can be made in respect of model [M6.9].

− Identifying the areas in which a Pareto-efficient DMU might prove an example of good operating practice for other DMUs to emulate. Virtual input and output levels of a Pareto efficient DMU can be used for this purpose under VRS in the same manner as they were under CRS.

− Identifying efficient DMUs whose operating practices a Pareto inefficient DMU may attempt to emulate to improve its performance.
Such DMUs are the efficient peers to an inefficient DMU. They are identified by the constraints to [M6.8] or [M6.9] which are binding at its optimal solution.

6.8   DEA ASSESSMENTS UNDER VRS BY MEANS OF *WARWICK DEA SOFTWARE*

DEA software can be used readily to ascertain the pure technical input or output efficiency of a DMU, and when it is Pareto-efficient, to identify the nature of returns to scale at its locality.

Example 6.6: Reconsider the 9 DMUs in Table 6.3 and maintain the assumption that they operate an input to output transformation process characterised by VRS. Use the *Warwick DEA Software* to ascertain the efficiency rating and the nature of returns to scale holding locally at DMU 6.

Solution: It is sufficient to print the table of the weights at an optimal solution to the value-based DEA model under VRS in respect of DMU 6 to derive the information required. To do so, first an input file is prepared, e.g. using the following format:
   -input1 −input 2 +output
   DMU1 10 15 10
   DMU2 15 10 10

   ...
   DMU9 30 27.5 20
   End.

Once the above input file has been read by the *Warwick DEA Software* the settings indicated on the screen in Figure 6.6 are used. The part of the log file containing the DEA weights for DMU 6 is as follows:

Virtual IOs for Unit DMU6 efficiency 100.00% radial				
OMEGA	-62.50%	-0.62500		-0.50000
VARIABLE	VIRTUAL IOs	IO WEIGHTS	NO LO BND	MAX
-INPUT1	50.00%	0.00833		0.01667
-INPUT2	50.00%	0.00833		0.00000
+ OUTPUT	162.50%	0.05417		0.05000

This shows that the efficiency rating of DMU 6 is 1 (100%). It also shows that one optimal solution (where the lowest virtual input or output is maximum) is for w = - 0.625 and weights of 0.00833 for each one of the inputs and 0.05417 for the output. (The interested reader can use these weights in model [M6.14] to verify that they are feasible and correspond to an efficiency rating of 1, normalising total virtual input to 1 rather than 100.)

The efficiency rating of 100% suggests that DMU 6 is Pareto-efficient. The label omega is used for the variable by that name in model [M6.8] or w in model [M6.9]. The software computes for each Pareto-efficient DMU the maximum and minimum omega value compatible with the maximum efficiency rating of the DMU. This gives a range of omega values which are compatible with the maximum efficiency rating of the DMU concerned. This range of omega values is used to identify the type of returns to scale holding locally at the DMU. The software has <u>adapted</u> the conditions in (6.8) and (6.9) to the following simple reporting form (see also Chapter 10):

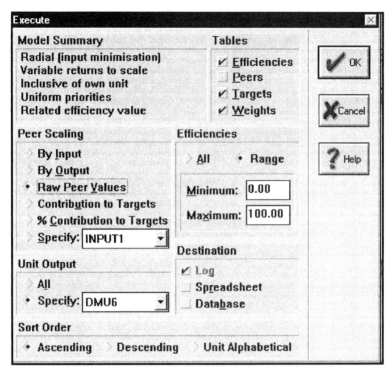

Figure 6.6.   Illustration of *Warwick DEA Software* screen for Assessments in the Input
Orientation under VRS

- If the range of omega values is positive the DMU concerned operates under increasing returns to scale;
- If the range of omega values includes zero the DMU concerned operates under CRS;
- If the range of omega values is negative the DMU concerned operates under decreasing returns to scale.

It should be noted that the foregoing statements are specific to the way *Warwick DEA Software* reports ranges to make easier the identification of the nature of returns to scale and NOT to be confused with the formal conditions contained in (6.8) and (6.9).

The output above shows that the maximum value of ω compatible with an efficiency rating of 100% for DMU 6 is -0.5 (the rest of the corresponding DEA weights being those appearing under that value of ω). Since the range of optimal omega values is negative (it has no lower bound) we conclude that at DMU 6 we have local DRS. This confirms our findings in Example 6.5.

6.9. DEA ASSESSMENTS UNDER NON-INCREASING OR NON-DECREASING RETURNS TO SCALE

It is possible for the DMUs being assessed to operate an input to output transformation process which is characterised by more specialised types of returns to scale than we have covered so far. Two such types of returns to scale are *non-increasing* and *non-decreasing* returns to scale. Figures 6.7 and 6.8 illustrate in a two-dimensional case the shape of the PPS boundary when it exhibits respectively non-increasing and non-decreasing returns to scale.

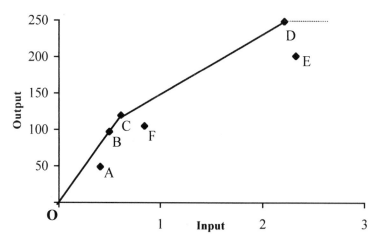

Figure 6.7 An Illustration of Non-Increasing Returns to Scale

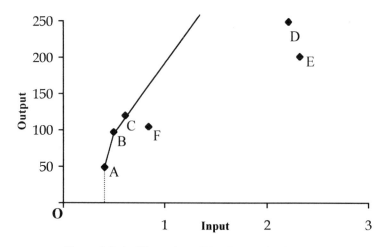

Figure 6.8. An Illustration of Non-Decreasing Returns to Scale

The envelopment models in [M6.1] and [M6.5] can be readily modified to assess efficiency under non-increasing or non-decreasing returns to scale. The modifications needed are as follows:

- To assess pure technical input (output) efficiency under non-increasing returns to scale replace the convexity constraint in model [M6.1] (model [M6.5]) by the constraint $\sum_{j=1}^{N} \lambda_j \leq 1$

$(\sum_{j=1}^{N} \alpha_j \leq 1)$.

- To assess pure technical input (output) efficiency under non-decreasing returns to scale replace the convexity constraint in model [M6.1] (model [M6.5]) by the constraint $\sum_{j=1}^{N} \lambda_j \geq 1$

$(\sum_{j=1}^{N} \alpha_j \geq 1)$.

For a further elaboration on the rationale of these modifications and a more *generalised returns to scale model* see Cooper et al. (2000) (section 5.7).

6.10 QUESTIONS

(Where the context permits the reader may employ the Warwick DEA Software accompanying this book to answer the questions below. See Chapter 10 for a limited User Guide.)

1.   A family owned pizza chain operates 8 outlets. Sales for the last 12 month-period by each outlet and pizza sales in total within the catchment area of each outlet are as follows:

Table 6.4.   Pizza Sales

Outlet	1	2	3	4	5	6	7	8
Outlet catchment Area Sales ($0,000)	4	6	8	2	5	7	8	9
Sales by the outlet ($000)	3	5	6	0.9	3.8	7.5	8	8.8

a)   Plot the production possibility set which can be used to assess the outlets on their marketing efficiency by DEA under VRS;
b)   Using the plot identify the Pareto-efficient outlets and the nature of the local returns to scale at each outlet;
c)   Identify whether input or output orientation would be appropriate in this case for assessing the pure technical efficiency of outlets and hence determine the pure technical efficiency of outlet 5;
d)   Using an envelopment DEA model assess the pure technical efficiency of outlet 7. What is the most productive scale size for the outlet? How can information on the most productive scale be of use in the context of the pizza outlets?
e)   Using a value-based DEA model identify the nature of local returns to scale where DMU 7 lies or is projected.

2.   Consider the six sales representatives in Chapter 5, Table 5.5. The data is reproduced below.

Representative	Hours of work (00)	Value of sales ($'000)	Households Signed Up
1	7.5	50	60
2	8	37	70
3	7	26	56
4	8.4	59	65
5	8.7	68	45
6	8.2	64	80

a) Assess the technical and the pure technical output efficiency of sales representative 3 using envelopment models. Hence compute and interpret her scale efficiency.

b) What is the most productive number of hours of work for salesperson 3 and which salesperson(s) is/are the role-model(s) underlying your statement?

c) Ascertain the type of returns to scale holding locally on the efficient boundary where sales person 3 is located or projected.

3.   You manage six retail outlets selling clothes and wish to assess them on their efficiency in attracting custom (*marketing efficiency*). You decide to use DEA so that you can judge the sales level of each store relative to the size of the market in which the store operates and the size of the floor area it has. Relevant data is as follows.

Table 6.5.   Clothes Sales Data

	Sales by the shop ($m)	Catchment area sales ($m)	Floor space (m²)
Shop 1	3.5	35	110
Shop 2	8	67	200
Shop 3	11	108	220
Shop 4	7	80	190
Shop 5	4	35	120
Shop 6	5	48	80

a) Identify other factors additional to those above which may need to be taken into account to judge the market efficiency of each shop. Justify each factor and explain the relevant measure you would use within the model.

b) Explain why it would be more sensible to assess the market efficiencies of the shops
   - using an output orientation and
   - using a VRS rather than a CRS DEA model.

c) In assessing Shop 4 using an envelopment VRS output maximising DEA model you find that it has shops 2, 3 and 6 as its efficient peers. All slack variables in the model are zero and the value of $\lambda 6$ (the variable relating to shop 6) is 0.148. Estimate the market efficiency of Shop 4 and interpret it.

4.   Maintaining a variable returns to scale assumption select a relatively inefficient DMU from those in Table 6.6.

a) Estimate the efficiency of the DMU given the scale of its outputs.
b) Hence estimate the scale efficiency of the DMU.
c) Estimate targets that would render the DMU most productive scale size. Contrast these with the targets that would eliminate managerial inefficiency at the current scale of operations and make comments on any differences between the two sets of targets.

Table 6.6. Input Output Data on Local Property Tax Offices

	COST	HEREDS	REBATES	SUMMONS	VALUE
DMU1	9.130	7.525	34.114	21.958	3.840
DMU2	13.600	8.301	23.270	35.966	8.632
DMU3	5.760	10.909	13.392	11.527	4.913
DMU4	11.240	16.621	36.817	27.552	7.522
DMU5	15.760	22.809	95.776	23.611	12.266
DMU6	5.650	1.777	0.156	1.314	39.011
DMU7	21.600	15.107	70.958	54.216	10.809
DMU8	8.570	7.919	48.688	14.032	5.923
DMU9	6.010	7.066	36.304	5.445	2.936
DMU10	8.020	8.858	43.610	13.774	4.274
DMU11	9.930	8.999	36.852	20.661	8.151
DMU12	7.900	8.278	45.222	6.191	5.327
DMU13	5.150	6.763	18.704	10.620	3.540
DMU14	6.070	8.984	19.980	14.860	4.060
DMU15	5.260	7.686	25.970	9.930	3.860
DMU16	7.366	10.640	23.428	17.596	6.274
DMU17	3.896	5.638	23.672	5.830	3.900
DMU18	10.167	14.310	30.540	17.770	8.005
DMU19	21.970	12.234	92.020	29.530	14.763
DMU20	9.700	7.674	41.162	13.272	4.503

Chapter 7

ASSESSING POLICY EFFECTIVENESS AND PRODUCTIVITY CHANGE USING DEA

7.1 INTRODUCTION

The notion of assessing policy effectiveness by means of DEA was first introduced by Charnes et al. (1981). At issue is the disentangling of *managerial* and *policy* efficiency. A simple example will illustrate what we mean here. The author was involved in an assessment of the "market efficiency" of a set of public houses (*pubs*) in England, part of which is reported in Athanassopoulos and Thanassoulis (1995). The market efficiency of an outlet was intended to reflect its ability to attract custom, controlling for its environment and other factors beyond managerial control. The pubs to be assessed could be clearly separated into two categories: Pubs located so as to benefit from passing trade and pubs aimed mostly at local residents. In the context of this chapter we would say that each category of pubs was operating under a different '*policy*' in terms of where the pubs were sited, the style of their management, internal décor and so on. Many real life contexts comprise operating units which perform similar tasks but could be said to operate under different policies. For example Golany and Storbeck (1999) in an assessment of bank branches identify those with "Personal Investment Centres" as one "programme" (policy in our context) and those without such centres as another programme or policy.

The first part of this chapter addresses the issue of how to disentangle any inefficiency identified at operating unit level into that attributable to the unit's management and that attributable to the policy under which the unit operates. This is achieved by decomposing a unit's efficiency into a component attributable to the management of the unit and a further component attributable to the policy under which the unit operates. The decomposition is multiplicative under certain circumstances.

The second part of the chapter deals with the use of DEA to assess *productivity change* over time. Assessment of productivity change by means of DEA was first introduced by Färe et al. (1989). In single-input single-output contexts the *productivity* of a unit is defined as the ratio of its output to its input. In multi-input multi-output contexts a unit's productivity is defined as

the ratio of an index of its output levels to an index of its input levels.

The change over time of this measure reflects the change in the unit's productivity.

Initially economists attributed productivity changes exclusively to technological changes, ignoring any impact on a unit's productivity due to changes in its efficiency as distinct from changes in the technology the unit operates. However, it is now accepted that productivity can change by a combination of so-called *boundary shift* and *efficiency change*. Boundary shift reflects improvement or regress in the industry efficient boundary which in turn reflects changes in the technology used by the units in the industry. Efficiency change on the other hand reflects an operating unit's own shift over time relative to the industry efficient boundary.

Färe et al. (1989) used DEA to compute a *Malmquist Index* which measures a unit's overall productivity change. The index is decomposed multiplicatively into a boundary shift and an efficiency change component. The Malmquist Index captures productivity change in terms of quantities without reference to input prices or output values. Maniadakis and Thanassoulis (2000 and forthcoming) developed a modification to the Malmquist Index to reflect productivity changes in terms of the combined effect of input costs or output values and physical quantities relating to the two time periods over which productivity change is being measured.

This chapter details the use of DEA to assess productivity change over time using decompositions of efficiency measures.

7.2. DISENTANGLING MANAGERIAL AND POLICY EFFICIENCY: AN OUTLINE

Where we have two or more groups of units performing the same function but under different policies we often want to know whether the units operating under one policy are intrinsically more effective than those operating under another. We may wish to have this information for the purpose of assessing retrospectively the effectiveness of those echelons of management who made the policy decisions and/or for the purpose of planning to institute more effective policies for the future. It is not straightforward to address the question of whether some policy intrinsically makes units more effective. If the units are assessed by policy group, the efficiency ratings are not comparable across groups since they relate to different benchmark units for each group. If all the units are assessed in a single group, the efficiency rating of each individual unit will reflect a combination of the performance of its management and of the impact of the policy under which the unit operates. The approach developed by Charnes et al. (1981) provides a way to disentangle managerial from policy effectiveness.

The approach involves a two-stage assessment process. In the first stage, the analyst assesses each unit within its own policy group.

The DEA efficiency rating of each unit within its policy group is referred to as its *managerial efficiency*.

The managerial efficiency of a unit reflects its performance "net" of any policy effects since the unit has been assessed with reference only to other units which operate the same policy as the unit itself. Any inefficiency cannot therefore be attributed to the policy under which the unit operates.

The first stage assessment makes it possible to estimate a set of input-output levels that would render the unit Pareto-efficient within its own policy group. The sets of input-output levels traditionally used in this respect and those we shall use here are as defined in Chapter 4, in (4.1) (input orientation) or alternatively (4.2) (output orientation). These input-output levels are referred to as *radial targets* because they reflect the attainment of Pareto-efficiency through pre-emptive priority to radial input contractions or output expansions.

The analyst now replaces the observed input-output levels of each unit by its radial targets, pools the units into a single group, and assesses them afresh. Any inefficiencies identified at this second-stage assessment are attributable to the policies within which the units operate rather than to their management. This is because the analyst has artificially "eliminated" managerial inefficiencies by adjusting the data of all units to within-policy efficient levels.

The DEA efficiency rating corresponding to the radial targets of a unit is referred to as the *policy efficiency* at the input-output mix of the unit concerned.

Figure 7.1 illustrates how the foregoing approach works. The figure shows a set of operating units which use two inputs to produce a single output. The units are assumed to operate a constant returns to scale technology and Figure 7.1 shows their input levels per unit of output. The units operate under two policies, those labelled * operate under policy 1 and those labelled + operate under policy 2.

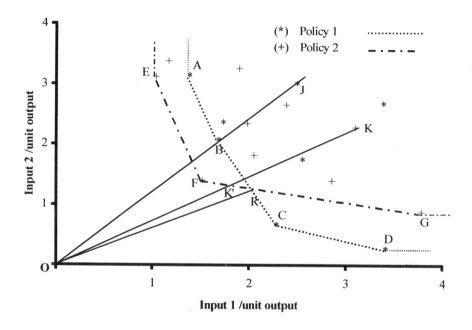

Figure 7.1.   Separating Managerial and Policy Efficiencies

ABCD is the efficient boundary for units operating under policy 1 and EFG for units operating under policy 2. In the stage 1 assessment each unit is assessed only within its own policy group. The efficiency rating obtained at this stage is attributable to the unit's management. For example

$$\text{the managerial efficiency of unit J is } \frac{OB}{OJ} \text{ and that of unit K } \frac{OK'}{OK}.$$

To assess now the component of each unit's inefficiency which can be attributed to the policy under which the unit operates, its input-output levels are adjusted to its radial targets as estimated within its policy group. Thus for example in Figure 7.1 the input levels of unit K are adjusted to those at K' which would render the unit Pareto-efficient within its parent policy while maintaining its input mix. The adjusted units to be assessed are now as in Figure 7.2.

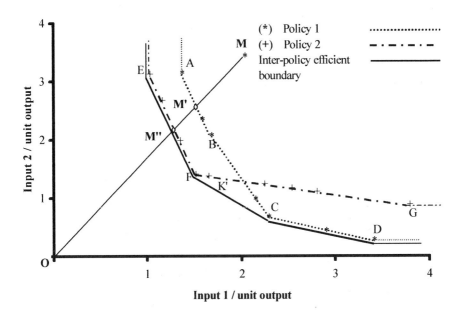

Figure 7.2. Assessing Policy Efficiency Using Adjusted Input Levels

The inter-policy efficient boundary is EFCD enveloping the *adjusted* units of the two policies. Any inefficiency of an adjusted unit relative to the inter-policy efficient boundary is now attributable to the policy under which the unit operates rather than to its management.

For example at the input mix of M the policy efficiency is $\dfrac{OM''}{OM'}$.

This is a measure of the component of the inefficiency of unit M which is attributable to the policy under which the unit operates. Note that relative to the inter-policy boundary unit M in Figure 7.2 has efficiency $\dfrac{OM''}{OM}$ while relative to the policy 1 boundary its efficiency is $\dfrac{OM'}{OM}$. Thus we have the following multiplicative decomposition:

$$\text{Inter-policy efficiency of M} = \frac{OM''}{OM} = \frac{OM'}{OM} \times \frac{OM''}{OM'} \qquad (7.1).$$

or

Inter-policy efficiency of M =
Managerial efficiency of M × Policy efficiency at the input mix of M.

This multiplicative decomposition will not be valid when any one of the efficiency components in (7.1) is measured relative to a reference point on the inefficient part of the boundary. Such parts are represented in Figure 7.2 by the vertical and horizontal extensions to the policy boundaries. In such a case the radial targets used in the two-stage optimisation procedure outlined would move the reference point from an inefficient to an efficient part of the boundary before the policy-efficiency measure is computed. Thus a multiplicative decomposition of the type depicted in (7.1) will no longer be valid.

7.3   DISENTANGLING MANAGERIAL AND POLICY EFFICIENCY: THE GENERIC APPROACH

Let us assume that we have N_p DMUs operating under policy p ($p = 1...P$). Let us further assume that DMU j of policy p uses inputs x_{ij}^p i = 1...m to secure outputs y_{rj}^p r = 1...s.

Stage 1: Assessing managerial efficiency

The managerial technical input efficiency of DMU j_0 of policy p is $k_{j_0}^{p*}$, where $k_{j_0}^{p*}$ is the optimal value of $k_{j_0}^p$ in model [M7.1].

Min $k_{j_0}^p - \varepsilon [\sum_{i=1}^{m} S_i^- + \sum_{r=1}^{s} S_r^+]$ [M7.1]

Subject to:

$$\sum_{j=1}^{N_p} \lambda_j x_{ij}^p = k_{j_0}^p x_{ij_0}^p - S_i^-   \qquad i = 1...m$$

$$\sum_{j=1}^{N_p} \lambda_j y_{rj}^p = S_r^+ + y_{rj_0}^p   \qquad r = 1...s$$

$\lambda_j \geq 0$, $j=1...N_p$, S_i^-, $S_r^+ \geq 0 \forall i and r$, $k_{j_0}^p$ free.

ε is a non-Archimedean infinitesimal.

This is the generic model for assessing technical input efficiency introduced in Chapter 4, (model [M4.1])) applied here over the comparative set of DMUs N_p operating under the same policy p as DMU j_0.

Stage 2: Assessing policy efficiency

Let $(x_{ij_0}^{tp}, i=1...m, y_{rj_0}^{tp}, r=1...s)$ be a set of input-output levels that would render DMU j_0 Pareto-efficient within its policy p. We will use the set $(x_{ij_0}^{tp}, i=1...m, y_{rj_0}^{tp}, r=1...s)$ yielded by model [M7.1] so that

$$x_{ij_0}^{tp} = \sum_{j=1}^{N_p} \lambda_j^* x_{ij}^p = k_{j_0}^{p*} x_{ij_0}^p - S_i^{-*} \qquad\qquad i = 1...m$$

$$y_{rj_0}^{tp} = \sum_{j=1}^{N_p} \lambda_j^* y_{rj}^p = S_r^{+*} + y_{rj_0}^p \qquad\qquad r = 1...s \qquad\qquad (7.2),$$

where the superscript * denotes the optimal value of the corresponding variable in [M7.1].

Set the input-output levels of all DMUs $j \in N_p$ of each policy $(p = 1...P)$ to their respective target input-output levels as defined in (7.2). (The issue of alternative input-output levels which can render a DMU Pareto-efficient within its own policy set is discussed later in this section.) The policy efficiency *at the input-mix* of DMU j_0 is p_0^*, the optimum value of p_0 in [M7.2]. This is the generic model for assessing technical input efficiency introduced in Chapter 4 (model [M4.1]) applied here over a comparative set created by pooling all DMUs of all policies together, after first adjusting their input-output levels to their radial targets as defined in (7.2).

Min $\qquad\qquad p_0 - \varepsilon[\sum_{i=1}^{m} S_i^- + \sum_{r=1}^{s} S_r^+]$ $\qquad\qquad$ [M7.2]

Subject to:

$$\sum_{p=1}^{P} \sum_{j=1}^{N_p} \lambda_j x_{ij}^{tp} = p_0 x_{ij_0}^{tp} - S_i^- \qquad\qquad i = 1...m$$

$$\sum_{p=1}^{P} \sum_{j=1}^{N_p} \lambda_j y_{rj}^{tp} = S_r^+ + y_{rj_0}^{tp} \qquad\qquad r = 1...s$$

$$\lambda_j \geq 0, \forall j, S_i^-, S_r^+ \geq 0 \; \forall i \text{ and } r, p_0 \text{ free.}$$

ε is a non-Archimedean infinitesimal.

Once the foregoing two-stage assessment has been implemented the relative effectiveness of some policy p can be judged by comparing the efficiencies p_0^* of the DMUs in policy p against those of the DMUs in other policies. For example summary measures such as mean, median and quartile values of the efficiencies p_0^* in policy p convey a measure of the distance between the policy p efficient boundary and the inter-policy efficient boundary. The greater this distance the less effective the DMUs of policy p and therefore the less effective is policy p intrinsically.

Comparisons of policies in the manner outlined here must be used with caution, however. Firstly, a policy may contain too few DMUs on which to base conclusions about its effectiveness. Secondly, it may be the case that a policy is not globally more effective than another, but rather that it is more effective under certain circumstances (operating contexts) and not others. This type of situation is discernible in Figure 7.1. Policy 1 is more effective for input 2 to input 1 ratios to the right and below that on OR while policy 2 is more effective for input 2 to input 1 ratios to the left and above OR. In multi-policy and multi-input multi-output scenarios the association, if any, between policy effectiveness and input or output mix can be complex and will need careful analysis to identify.

An important point to note is that the policy efficiency measure model [M7.2] yields with reference to DMU j_0 depends on the particular point on the efficient boundary we reflect that DMU at the end of the first stage assessment. We have used so far the '*radial*' targets defined in (7.2) to reflect each DMU on a point of its policy's efficient boundary corresponding to giving pre-emptive priority to the radial improvement of input or alternatively output levels. However, model [M7.1] may well have alternative optimal solutions leading to alternative radial targets for DMU j_0. Moreover, as we will see later in Chapter 9, we may choose to estimate non-radial targets for each DMU. We can in fact choose to reflect, in principle, an inefficient DMU anywhere on the efficient boundary of its policy if we choose suitable priorities of improvement for inputs and outputs. (The *Warwick DEA Software* incorporates elements of such a facility (see in Chapter 10).)

If non-radial targets are used in the context of the two-stage procedure outlined here then policies may be compared only on a subset of the input (or output) mixes found within the DMUs operating under each policy. For example in Figure 7.1 if all DMUs are reflected on the efficient parts of the policy boundaries lying to the right and below OR then Policies 1 and 2 will only be compared for DMUs operating the input mixes to the right and below that on OR.

In summary, comparison of policies on intrinsic efficiency needs to be designed to capture all the input-output mixes on which the user wishes to compare the policies.

7.4   DISENTANGLING MANAGERIAL AND POLICY EFFICIENCY: ILLUSTRATIVE EXAMPLES

Example 7.1: Reconsider the six distribution centres of Example 6.1, Chapter 6. Assume that the centres operate under constant returns to scale and that Centres 1-3 inclusive lease the delivery vehicles they use (henceforth *Lease Policy*), while the rest of the centres own their delivery vehicles (henceforth *Buy Policy*).

a)   Using linear programming compute the managerial efficiency of Centre 1 and the policy efficiency at its output mix.

b)   Repeat (a) using the *Warwick DEA Software*.

Solution: a) To assess Centre 1 within the Lease Policy centres 4-5 are ignored and the following linear programming model is solved:

Min $Z - \varepsilon (S1 + S2 + S3)$ [M7.3]
Subject to:

$$15\lambda_1 + 14.5\lambda_2 + 16.5\lambda_3 - S1 = 15$$
$$10.5\ \lambda_1 + 12\lambda_2 + 23\lambda_3 - S2 = 10.5$$
$$38\lambda_1 + 36\lambda_2 + 42\lambda_3 + S3 = 38Z$$
$$S1, S2, S3, \lambda_1, \lambda_2, \lambda_3 \geq 0.$$

The optimal solution yields $Z = 0.98$, $\lambda_2 = 1.034483$ and all other λ's zero. Thus the managerial efficiency of Centre 1 is 0.98.

To compute the policy efficiency at the output mix of Centre 1 it is necessary first to identify the input-output levels which will render each one of the six centres Pareto-efficient within its own policy. Then the centres must be grouped using these Pareto-efficient input-output levels and Centre 1 assessed within the aggregate set.

The targets that will render Centre 1 Pareto-efficient within the Lease Policy can be computed using the optimal λ values of model [M7.3] within the generic formula (4.1) in Chapter 4. They are:
(Tonnes Delivered, Customer points served, Operating expenditure) =
$\lambda_2 \times$ Centre 2 = 1.034483 (14.5,12,36) = (15, 12.41,37.24).

Since Centre 2 was an efficient peer for Centre 1 above we know that its observed input-output levels render it Pareto-efficient and no need arises to modify and solve model [M7.3] in respect of Centre 2. When model [M7.3] is modified and solved in respect of Centre 3 it too turns out to be Pareto-efficient within the Lease Policy.

To identify now the targets that will render Centre 4 Pareto-efficient within the Buy Policy the following model is solved:

Min $Z - \varepsilon (S1 + S2 + S3)$ [M7.4]
Subject to:
$$18\lambda_4 + 21\lambda_5 + 11\lambda_6 - S1 = 18$$
$$21\lambda_4 + 15\lambda_5 + 8 \lambda_6 - S2 = 21$$
$$45\lambda_4 + 37\lambda_5 + 17\lambda_6 + S3 = 45Z$$
$$S1, S2, S3, \lambda_4, \lambda_5, \lambda_6 \geq 0.$$

The optimal solution yields $Z = 0.99$, $\lambda_6 = 2.625$ and all other λ's zero.
Thus the managerial efficiency of Centre 4 is 0.99 and the targets that will render it Pareto-efficient within the Buy Policy are
(Tonnes Delivered, Customer points served, Operating expenditure) =
$2.625 (11,8,17) = (28.875, 21, 44.625)$.
Clearly the observed input-output levels of Centre 6 are Pareto-efficient within the Buy Policy. Modifying and solving model [M7.4] yields the following Pareto-efficient input-output levels for Centre 5:
(Tonnes Delivered, Customer points served, Operating expenditure) =
$(21,15.27,32.45)$.

To assesses now the Policy efficiency at the output mix of Centre 1 we combine all six centres at their within-policy Pareto efficient input-output levels yielding the following model:

Min $Z - \varepsilon (S1 + S2 + S3)$ [M7.5]
Subject to:
$$15\lambda_1 + 14.5\lambda_2 + 16.5\lambda_3 + 28.875\lambda_4 + 21\lambda_5 + 11\lambda_6 - S1 = 15$$
$$12.41 \lambda_1 + 12\lambda_2 + 23\lambda_3 + 21\lambda_4 + 15.27\lambda_5 + 8 \lambda_6 - S2 = 12.41$$
$$37.24\lambda_1 + 36\lambda_2 + 42\lambda_3 + 44.625\lambda_4 + 32.45\lambda_5 + 17\lambda_6 + S3 = 37.24Z$$
$$S1, S2, S3, \lambda_1, \lambda_2, \lambda_3, \lambda_4, \lambda_5, \lambda_6 \geq 0.$$

The efficiency yielded by model [M7.5] in respect of 'Centre 1' is 0.683. Thus the Policy Efficiency at the output mix of Centre 1 is 68.3%. (Strictly speaking this policy efficiency is at the output mix that renders Centre 1 Pareto-efficient which on this occasion is identical to its observed output mix.)

This means that the management of Centre 1 is doing quite well within the Lease Policy that they operate. They are 98% efficient within their policy. However, the policy itself is not particularly effective, at least not for their output mix. The Centre with its current output mix would only be

68.3% efficient relative to all six centres even if it had reached the Pareto efficient boundary of the Lease Policy.

b) The foregoing computations can be performed using the *Warwick DEA Software*. The input file is

```
+TONNES +CUSTOMERS -OPEX
C1 15 10.5 38
C2 14.5 12 36
C3 16.5 23 42
C4 18 21 45
C5 21 15 37
C6 11 8 17
End.
```

The input file is read by the software and then under *Options* the default setting is used (i.e. *input minimising, radial model, constant returns to scale, uniform priorities, inclusive of own unit*). To assess efficiencies and targets under the Lease Policy centres 4-5 are taken out of the active list, by means of *Run* menu, *Select Units*. See the screen in Figure 7.3. Tick *efficiencies* and *targets* under *Execute* in the *Run* menu and do the same for *Spreadsheet* under *Destination* (of output). The efficiencies file produced shows the efficiency of 98% for Centre 1 matching that computed by linear programming earlier. The *targets* file produced by the software includes the observed and targets levels. The targets correspond to the columns labelled _T. They are reproduced below and match the targets we computed for Centres 1-3 above using model [M7.3].

```
UNIT, OPEX_T,TONNES_T, CUSTOMERS_T
C1, 37.2, 15.0, 12.4
C2, 36.0, 14.5, 12.0
C3, 42.0, 16.5, 23.0
```

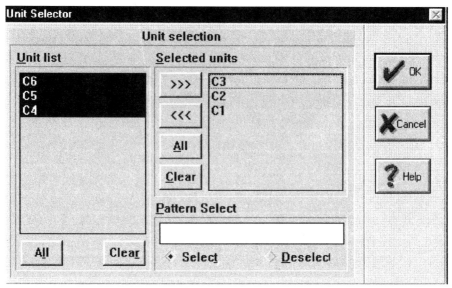

Figure 7.3. Centres C1-C3 Constitute the Active Units (DMUs)

To identify Pareto-efficient targets under the Buy Policy Centres 4-5 are moved to the active list and Centres 1-3 are moved to the inactive list. This is done using *Select Units* under the *Run* menu as above. Then the run is effected to produce the following Pareto-efficient targets.

UNIT, OPEX_T, TONNES_T, CUSTOMERS_T
C4, 44.6, 28.9, 21.0
C5, 32.5, 21.0, 15.3
C6, 17.0, 11.0, 8.0

Combining the targets under both policies the following input file is created.
   -OPEXT +TONST +CUSTT
   C1 37.2 15.0 12.4
   C2 36.0 14.5 12.0
   C3 42.0 16.5 23.0
   C4 44.6 28.9 21.0
   C5 32.5 21.0 15.3
   C6 17.0 11.0 8.0
   End.

Using this input file Centre 1 is assessed afresh. Its efficiency rating is 68.29% as determined above using linear programming. This is
   the Lease Policy efficiency at the output mix of Centre 1.

Example 7.2: Consider the local tax offices (rates departments) whose data appears in Appendix 5.2. Assume that the first 31 departments operate in urban areas while the rest operate in rural areas. Pay scales for staff are at the same levels for rural and urban areas. Assume the departments operate under variable returns to scale and that output levels are exogenous.

a) Identify the component of the inefficiency of each department attributable to its management;

b) Investigate whether there is any evidence that urban departments are intrinsically more efficient than their rural counterparts.

Solution: a) Managerial inefficiency at each department is identified by assessing it within its own policy. In this manner we control for department type (urban or rural) and assess managerial performance. Using "Cost" as input and the rest of the variables in Appendix 5.2 as outputs an input-oriented variable returns to scale DEA model is solved separately within the urban and rural sets of departments. The efficiencies in Table 7.1 are obtained.

b) To identify whether urban or rural departments are intrinsically more efficient we first "eliminate" managerial inefficiency and then assess the departments. Managerial inefficiency is eliminated by adjusting the input-output levels of the departments to those which would render them Pareto-efficient within their respective policies. Such input-output levels are defined in (7.2) and have been computed here using the *Warwick DEA Software.* They can be found in Appendix 7.1.

Using now an input oriented DEA model on the combined set of 62 departments with the input-output levels of Appendix 7.1 the efficiencies in Table 7.2 are obtained. It is clear by inspection that the rural departments are less efficient than the urban departments. (It must be recalled here that for the purposes of this illustrative exercise the split of what were in fact homogeneous departments into urban and rural departments was arbitrary.)

7.5   ASSESSING PRODUCTIVITY CHANGE BY MEANS OF DEA: A GRAPHICAL INTRODUCTION TO THE MALMQUIST INDEX

As noted earlier we can define the productivity of a unit in the single-input single-output case as the ratio of its output to its input level. In the multi-input multi-output case this notion of average product breaks down. An approach has, however, been developed for measuring productivity change in such cases by drawing on the notion of indices as used in

DATA ENVELOPMENT ANALYSIS

Table 7.1. Managerial Efficiencies (%)

Urban	Efficiency	Rural	Efficiency
DEPT1	100	DEPT32	100
DEPT2	100	DEPT33	100
DEPT3	100	DEPT34	100
DEPT4	100	DEPT35	100
DEPT5	100	DEPT36	100
DEPT6	100	DEPT37	100
DEPT7	100	DEPT38	100
DEPT8	100	DEPT39	100
DEPT9	100	DEPT40	100
DEPT10	100	DEPT41	96.48
DEPT11	96.44	DEPT42	100
DEPT12	96.26	DEPT43	100
DEPT13	100	DEPT44	100
DEPT14	91.24	DEPT45	93.32
DEPT15	95.39	DEPT46	98.04
DEPT16	87.65	DEPT47	86.78
DEPT17	100	DEPT48	97.1
DEPT18	80.81	DEPT49	100
DEPT19	100	DEPT50	97.68
DEPT20	79.52	DEPT51	100
DEPT21	85.35	DEPT52	84.85
DEPT22	84.27	DEPT53	81.87
DEPT23	90.49	DEPT54	100
DEPT24	99.49	DEPT55	79.2
DEPT25	87.01	DEPT56	77.34
DEPT26	77.16	DEPT57	73.87
DEPT27	75.43	DEPT58	71.05
DEPT28	81.73	DEPT59	70.66
DEPT29	70.69	DEPT60	72.42
DEPT30	77.07	DEPT61	56.02
DEPT31	72.37	DEPT62	56.72

consumer theory. Specifically, Färe et al. (1989) have used DEA to compute a *Malmquist Index* of productivity change. They allowed for the fact that productivity change may be due to a combination of industry-wide productivity change over time and efficiency change at the level of the operating unit. They decomposed the index to capture these two components. Later, Färe et al. (1994b) decomposed the efficiency change component of the Malmquist Index into a pure technical and a scale efficiency change component.

Table 7.2.   Policy Efficiencies (%)

Urban	Efficiency	Rural	Efficiency
DEPT1	100	DEPT32	71.58
DEPT2	100	DEPT33	72.98
DEPT3	100	DEPT34	78.08
DEPT4	100	DEPT35	71.99
DEPT5	100	DEPT36	72.26
DEPT6	100	DEPT37	68.74
DEPT7	100	DEPT38	66.01
DEPT8	100	DEPT39	83.46
DEPT9	100	DEPT40	93.42
DEPT10	100	DEPT41	73.04
DEPT11	99.81	DEPT42	64.1
DEPT12	100	DEPT43	60.81
DEPT13	100	DEPT44	67.58
DEPT14	99.64	DEPT45	65.2
DEPT15	99.35	DEPT46	59.53
DEPT16	100	DEPT47	73.13
DEPT17	100	DEPT48	61.62
DEPT18	100	DEPT49	58.41
DEPT19	92.06	DEPT50	67.6
DEPT20	100	DEPT51	100
DEPT21	100	DEPT52	66.73
DEPT22	100	DEPT53	64.27
DEPT23	100	DEPT54	88.47
DEPT24	100	DEPT55	69.91
DEPT25	100	DEPT56	77.75
DEPT26	100	DEPT57	82.63
DEPT27	100	DEPT58	67.45
DEPT28	100	DEPT59	73.84
DEPT29	100	DEPT60	69.53
DEPT30	100	DEPT61	73.15
DEPT31	100	DEPT62	82.13

The Malmquist Index can be computed in the input orientation, controlling for output levels and measuring changes in input use, or alternatively in the in the output orientation, controlling for input use and estimating output level changes. However,

the DEA efficiencies needed are computed maintaining a constant returns to scale assumption *irrespective* of the actual returns to scale characterising efficient production in the technology operated by the units being assessed.

This way all productivity changes, including those attributable to changes in scale size over time are captured in the index. (For more on this point see Färe and Grosskopf 1994.)

As the Malmquist Index is always computed maintaining a constant returns to scale assumption, its value is the same whether it is computed in the input or in the output orientation. Hence, to simplify matters, we shall use the *input orientation.*

We can illustrate graphically how the Malmquist index measures productivity change. Consider a set of companies engaged in water distribution. Assume further that the output is the *amount of water delivered* (Megalitres) and the inputs are two: *operating expenditure* (OPEX) and *capital expenditure* (CAPEX). Finally assume that efficient operation is characterised by constant returns to scale and all companies face the same labour, resource etc. prices. (These are undoubtedly simplifying assumptions but maintaining them makes the illustration easier.)

We can think of productivity in this context as the ratio of the amount of water delivered to some index of input use by a company. Let Figure 7.4 depict the input levels OPEX and CAPEX per unit of output (Megalitre) at the companies shown. The efficient boundary is shown and to simplify matters we shall assume that it has NOT moved between period t and t+1. **This is equivalent to assuming that there has been no change in productivity at the <u>industry level</u> between period t and t+1.** Nevertheless, at company level there might be productivity gain or loss by virtue of the fact that a company has become more or less efficient over time.

In this regard, let some company operate at point F in period t and at point G in period t+1, Figure 7.4. We wish to know whether the company's productivity has improved or worsened over time.

The company's productivity change is measured by the ratio of its period t+1 to its period t efficiency measure.

In the context of Figure 7.4 the technical input efficiency of company F is $\dfrac{OD}{OF}$ in period t and $\dfrac{OE}{OG}$ in period t+1. Thus its productivity change is $\dfrac{OE}{OG} \div \dfrac{OD}{OF}$. We can see by inspection that this ratio is less than 1 and so the company's productivity is lower in period t+1 than in t. The company is further from the efficient boundary in period t+1 than it was in t while the boundary itself has not moved.

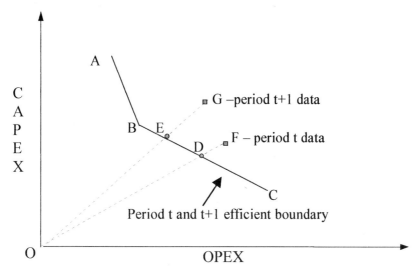

Figure 7.4.   Measuring Productivity Change

We can see fairly easily that had the ratio $\dfrac{OE}{OG} \div \dfrac{OD}{OF}$ above been 1 it will have meant that the company has stayed the same distance from the boundary in periods t and t+1 and since the boundary did not move the company has not changed productivity. Finally, if the foregoing ratio had been above 1 we could have readily concluded that the company experienced productivity progress because it is closer to the stationary efficient boundary in period t+1 than it was in period t. (Note that some authors in this field compute the Malmquist index by taking the ratio of period t to period t+1 efficiencies. This inverts the ratios we compute above and so we have productivity progress when the ratio is under 1 and so on.)

Let us now take the more realistic case where not only companies move relative to the efficient boundary over time but the boundary itself also moves. The situation that could arise is depicted in Figure 7.5.

To measure productivity change in Figure 7.5 we still use as in Figure 7.4 the ratio of the company's technical input efficiency in period t+1 to that in period t. However, we now compute this ratio once relative to the t period boundary and once relative to the t+1 period boundary. The geometric mean of these two ratios is a measure of the productivity change of the company.

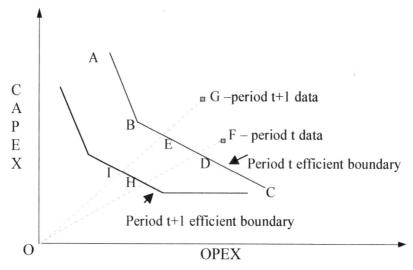

Figure 7.5. Measuring Productivity Change when the Efficient Boundary Moves Over Time

Thus the productivity change of the company operating at F in period t and at G in period t+1 in Figure 7.5 is

$$\{[\frac{OE}{OG} \div \frac{OD}{OF}] \times [\frac{OI}{OG} \div \frac{OH}{OF}]\}^{0.5} \tag{7.3}.$$

This geometric mean is company F's *Malmquist* index of productivity change between period t and t+1.

After some algebraic manipulation this index can be decomposed as follows.

Table 7.3.   Decomposition of the Malmquist Index of the Productivity Change of the Company Operating at F in Period t and at G in Period t+1, Figure 7.5.

$\{[\frac{OE}{OG} \div \frac{OD}{OF}] \times [\frac{OI}{OG} \div \frac{OH}{OF}]\}^{0.5} =$	$[\frac{OI}{OG} \div \frac{OD}{OF}] \times$	$[\frac{OE}{OI} \frac{OD}{OH}]^{0.5}$
↑	↑	↑
Malmquist index of productivity change of company F	'Catch-up' component	'Boundary shift' component

The *catch-up* term in Table 7.3 is a measure of how much closer to the boundary the company is in period t+1 compared to period t. This can be seen quite easily in Figure 7.5 where $\frac{OI}{OG}$ measures the distance of the company from the t+1 efficient boundary and $\dfrac{OD}{OF}$ its distance from the efficient boundary in period t. If the catch up term is 1 the company has the same distance in periods t+1 and t from the respective efficient boundaries. If the catch up term is over 1 the company has moved closer to the period t+1 boundary than it was to the period t boundary and the converse is the case if the catch up term is under 1.

The **boundary shift** term measures the movement of the boundary between periods t and t+1 at two locations: The ratio $\dfrac{OE}{OI}$ measures the distance of the two boundaries at the input mix of the company in period t+1. The ratio $\dfrac{OD}{OH}$ measures the distance of the two boundaries at the input mix of the same company in period t. The boundary shift is the geometric mean of these distances. Clearly in Figure 7.5 the boundary shift term is over 1 because both of the foregoing ratios are over 1. This represents productivity gain by the industry in that at both input mixes considered we have lower input levels in period t+1 than in period t for the standard unit of output. We can see quite easily that had the boundary shift term been 1 either the t and t+1 boundaries would have been coincident at the two input mixes involved, or at one input mix inputs in period t+1 would have exceeded those in period t and the converse would have been true at the other so that 'on average' the industry boundary would have shown neither gain nor loss in productivity. Finally a boundary shift term under 1 would signal that the industry has registered productivity loss as input levels would on average be higher in period t+1 compared to period t, controlling for output.

7.6   ASSESSING PRODUCTIVITY CHANGE BY MEANS OF DEA: THE MALMQUIST INDEX IN THE GENERAL CASE

Let DMUs $(j = 1...N)$ operate in periods t and t+1, the jth one using in period t inputs x_{ij}^t $(i = 1...m)$ to secure outputs y_{rj}^t $(r = 1...s)$. The Malmquist index of DMU j_0 in the input orientation is defined in (7.4).

$$MI_{j_0} = \left[\frac{C\_EF_{Tt}^{Dt+1} \; C\_EF_{Tt+1}^{Dt+1}}{C\_EF_{Tt}^{Dt} \; C\_EF_{Tt+1}^{Dt}} \right]^{1/2} \qquad (7.4),$$

where $C \_ EF_{Tt}^{Dt}$ is the (*radial*) technical input efficiency of DMU j_0 computed using its data of period t (Dt), relative to the efficient boundary (technology) of period t (Tt). The prefix "C_" indicates that the DEA efficiencies are computed maintaining a constant returns to scale assumption *irrespective* of the actual returns to scale characterising efficient production in the technology operated by the units being assessed. See the rationale for this in Färe and Grosskopf 1994 and the comment in the preceding section that **input and output oriented Malmquist indices are equal**.

The radial technical input efficiencies $C \_ EF_{Tt}^{Dt}$ in [7.4] are computed using model [M7.6].

Min k_0 [M7.6]
Subject to:

$$\sum_{j=1}^{N} \lambda_j x_{ij}^t - k_0 x_{ij_0}^t \leq 0 \qquad i = 1...m$$

$$\sum_{j=1}^{N} \lambda_j y_{rj}^t \geq y_{rj_0}^t \qquad r = 1...s$$

$\lambda_j \geq 0, j=1...N \geq 0$, k_0 free.

Thus the radial technical input efficiency $C \_ EF_{Tt}^{Dt}$ of DMU j_0 is $C \_ EF_{Tt}^{Dt} = k_0^*$, where k_0^* is the optimal value of k_0 in [M7.6]. The cross-time period radial technical input efficiencies are computed using models such as [M7.7].

Min q_0 [M7.7]
Subject to:

$$\sum_{j=1}^{N} \lambda_j x_{ij}^t - q_0 x_{ij_0}^{t+1} \leq 0 \qquad i = 1...m$$

$$\sum_{j=1}^{N} \lambda_j y_{rj}^t \geq y_{rj_0}^{t+1} \qquad r = 1...s$$

$\lambda_j \geq 0, j=1...N \geq 0$, q_0 free.

Model [M7.7] yields the cross-time period radial technical input efficiency $C \_ EF_{Tt}^{Dt+1}$ of DMU j_0 so that $C \_ EF_{Tt}^{Dt+1} = q_0^*$, where q_0^* is the optimal value of q_0 in model [M7.7]. The model can be readily modified to compute $C\_EF \ _{Tt+1}^{Dt}$ of DMU j_0.

The models in [M7.6] and [M7.7] ignore the slacks found in the generic input oriented DEA model [M4.1]. This in turn means that any productivity gain or loss which is not reflected in the *radial* DEA efficiency measures we compute with models [M7.6] and [M7.7] will not be captured by the Malmquist index. This has led to some criticism of the index and to the proposition of alternative measures of productivity change. (E.g. see Thrall (2000).)

After algebraic manipulation the Malmquist Index in (7.4) is decomposed as follows:

Table 7.4.   Decomposition of the Malmquist Index of DMU j_0

$$MI_{j_0} = \frac{C\_EF_{Tt+1}^{Dt+1}}{C\_EF_{Tt}^{Dt}} \times \left[\frac{C\_EF_{Tt}^{Dt+1}}{C\_EF_{Tt+1}^{Dt+1}} \times \frac{C\_EF_{Tt}^{Dt}}{C\_EF_{Tt+1}^{Dt}} \right]^{1/2}$$

↑	↑	↑
Index	'Catch-up' component	'Boundary shift' component

Notation in Table 7.4 is as in expression (7.4).

The ***catch-up*** term compares the closeness of DMU j_0 in each period to that period's efficient boundary. A value of 1 for this term would mean DMU j_0 has the same distance from the respective boundaries in periods t and t+1. A value of over 1 would mean DMU j_0 has become more efficient in period t+1 compared to period t, i.e. it has moved closer to the boundary in period t+1. Finally the converse is true when the catch-up term has a value under 1.

The interpretation of the ***boundary shift*** term in Table 7.4 is as in Table 7.3. A boundary shift in excess of 1 represents productivity gain by the industry in that at the input-output mixes of DMU j_0 in periods t and t+1 *efficient production* uses 'on balance' lower input levels in period t+1 than in period t, controlling for output levels. Note that efficient production in this context is as exhibited by the 'industry' and not necessarily by DMU j_0 itself. Similarly, a boundary shift under 1 represents productivity loss by the industry in that at the input-output mixes operated by DMU j_0 in periods t and t+1 *efficient production* uses 'on balance' higher input levels in period t+1 than in period t, controlling for output levels. Finally it follows from the foregoing that when the boundary shift term is 1 the industry has on average registered neither a productivity gain nor a productivity loss between period t and t+1. That is at the input-output mixes operated by DMU j_0 in periods t and t+1 *efficient production* uses 'on balance' the same input levels in period t+1 as in period t, controlling for output levels. (The term 'on balance' here gives expression to the geometric mean used for computing the distance between the efficient boundaries in periods t and t+1 at the two input-output mixes operated by DMU j_0 in those periods.)

7.7 ASSESSING PRODUCTIVITY CHANGE USING MALMQUIST INDICES: ILLUSTRATIVE EXAMPLE

Example 7.3: Reconsider the six distribution centres of Example 6.1, Chapter 6 but assume that the data covers three centres over two years as follows.

Year	Centre	Tonnes delivered (000)	Delivery points served (00)	Operating expenditure ($0,000)
1	1	15	10.5	38
1	2	14.5	12	36
1	3	16.5	23	42
2	1	18	21	45
2	2	21	15	37
2	3	11	8	17

a) Using linear programming compute the Malmquist Productivity Index of Centre 1 between years 1 and 2.
b) Decompose the Malmquist index of part (a) into an efficiency catch-up component and a boundary shift component.
c) Repeat (a) using the *Warwick DEA Software*.

Solution: a) To compute the Malmquist index in respect of Centre 1 we make recourse to the generic formula in (7.4) above which in the case of Centre 1 can be written as follows:

$$MI_1 = \left[\frac{C\_EF_{T1}^{D2}}{C\_EF_{T1}^{D1}} \frac{C\_EF_{T2}^{D2}}{C\_EF_{T2}^{D1}} \right]^{1/2}$$

where $C\_EF_{Tt}^{Dk}$ is the radial technical input efficiency of Centre 1 computed using its data of period k (Dk), relative to the efficient boundary (technology) of period t (Tt). The prefix "C_" indicates that the DEA efficiencies are computed maintaining a constant returns to scale assumption. To compute $C\_EF_{T1}^{D1}$ use model [M7.6] in respect of Centre 1 within centres 1-3 in year 1. The resulting linear programming model is [M7.8].

$$k_0^* = \text{Min } Z \hspace{4cm} \text{[M7.8]}$$

Subject to:

$$15\lambda_1 + 14.5\lambda_2 + 16.5\lambda_3 \geq 15$$
$$10.5\,\lambda_1 + 12\lambda_2 + 23\lambda_3 \geq 10.5$$
$$38\lambda_1 + 36\lambda_2 + 42\lambda_3 \leq 38Z$$
$$Z \text{ free, } \lambda_1, \lambda_2, \lambda_3 \geq 0.$$

This is yields $C\_EF_{T1}^{D1} = k_0^* = 0.98$.

To compute $C\_EF_{T2}^{D1}$ we use model [M7.7] setting it up in respect of Centre 1 within centres 1-3, year 2 data. The resulting linear programming model is as follows:

$$q_0^* = \text{Min } Z \hspace{4cm} \text{[M7.9]}$$

Subject to:

$$18\lambda_4 + 21\lambda_5 + 11\lambda_6 \geq 15$$
$$21\lambda_4 + 15\lambda_5 + 8\,\lambda_6 \geq 10.5$$
$$45\lambda_4 + 37\lambda_5 + 17\lambda_6 \leq 38Z$$
$$Z \text{ free, } \lambda_4, \lambda_5, \lambda_6 \geq 0.$$

The model yields $C\_EF_{T2}^{D1} = 0.61$.

To compute $C\_EF_{T1}^{D2}$ model [M7.8] above is modified using Centre 1 year 2 data within centres 1-3, year 1 data. The resulting linear programming model is as follows:

$$q_0^* = \text{Min } Z \hspace{4cm} \text{[M7.10]}$$

Subject to:

$$15\lambda_1 + 14.5\lambda_2 + 16.5\lambda_3 \geq 18$$
$$10.5\,\lambda_1 + 12\lambda_2 + 23\lambda_3 \geq 21$$
$$38\lambda_1 + 36\lambda_2 + 42\lambda_3 \leq 45Z$$
$$Z \text{ free, } \lambda_1, \lambda_2, \lambda_3 \geq 0.$$

The solution to this model yields $C\_EF_{T1}^{D2} = 1$.

Finally to compute $C\_EF_{T2}^{D2}$ model [M7.9] is set up in respect of Centre 1 year 2 data within centres 1-3, year 2 data. The resulting solution is $C\_EF_{T2}^{D2}$ =0.99.

Thus the Malmquist index reflecting the productivity change of Centre 1 between years 1 and 2 is

$$MI_1 = \left[\frac{C\_EF_{T1}^{D2}}{C\_EF_{T1}^{D1}} \frac{C\_EF_{T2}^{D2}}{C\_EF_{T2}^{D1}} \right]^{1/2} = \left[\frac{1}{0.98} \frac{0.99}{0.61} \right]^{1/2} = 1.2868.$$

This index value suggests Centre 1 has had productivity improvement of the order of 29% in the sense that its operating expenditure would be some 29% higher in year 1 compared to year 2 if we control for its output levels.

b) The generic formula for decomposing the Malmquist Index into an efficiency 'catch-up' and a 'boundary shift' component appears in Table 7.4. In the case of Centre 1 we have

'Efficiency catch-up' = $C\_EF_{T2}^{D2} / C\_EF_{T1}^{D1}$ = 0.99/ 0.98 =1.01.

We can deduce from Table 7.4 that:

Boundary Shift = Malmquist Index / Efficiency Catch-up = 1.2868 / 1.01 = 1.27.

Thus, Centre 1 has maintained a virtually constant distance in the two years from the respective efficient boundaries, as can be deduced from its efficiency catch-up component of 1.01. Virtually all its productivity gain is derived from the fact that the efficient boundary has become on average more productive in year 2 compared to year 1, at least at the output mixes operated by the Centre in the two years.

c) To compute the efficiencies needed in part (a) by means of the *Warwick DEA Software* we use its features which enable us to move DMUs (Centres) from the active to the inactive list and its facility to 'exclude' an active DMU as a potential efficient peer. (See Chapter 10 for more details on how these two features work.)

We begin by reading into the software data for the six DMUs. Then we make the year 2 data inactive (see Figure 7.6) and compute the efficiency of Centre 1 on year 1 data (technology). The result is $C\_EF_{T1}^{D1}$ = 0.98.

Then we select year 2 data and make year 1 data inactive to compute the efficiency of Centre 2 in year 2 on year 2 data. The result is $C\_EF_{T2}^{D2}$ = 0.99. To compute now the efficiency of Centre 1 with its year 2 data on the year 1 technology (comparators) we select all year 1 data plus the data of Centre 1 in year 2. See Figure 7.7 top for the relevant screen. We use the option to 'exclude' the unit being assessed (i.e. Centre 1 year 2 data) as a candidate efficient peer. See Figure 7.7 bottom for the 'execute' screen used. The result is $C\_EF_{T1}^{D2}$ = 1.00.

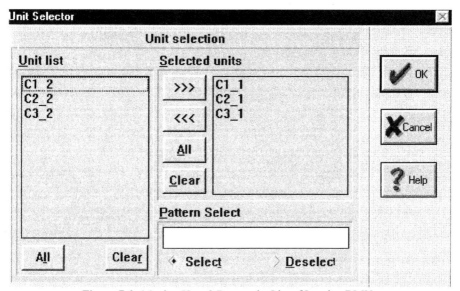

Figure 7.6. Moving Year 2 Data to the List of Inactive DMUs

The foregoing procedure is now repeated, this time to compute the efficiency of Centre 1 year 1 data on year 2 comparators. The result is $C\_EF_{T2}^{D1}$ = 0.61.

We have computed by *Warwick DEA Software* the four DEA efficiency figures needed for computing the Malmquist Index of productivity change of Centre 1 between Years 1 and 2. The efficiencies match those computed in part (a) using linear programming. Thus the Malmquist index and its components can also be derived using the *Warwick DEA Software*. (A batch mode of fuller versions of *Warwick DEA Software* automates the computation of cross-period efficiencies illustrated here.)

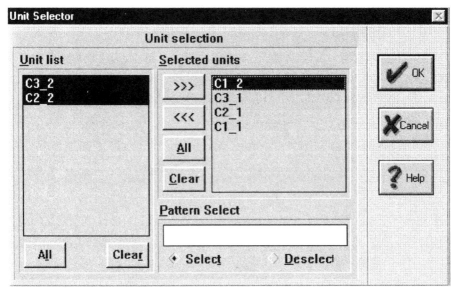

Figure 7.7. Selecting All Year 1 Data Plus Centre 1 Year 2 Data (above) and, Excluding the Assessed Unit (Centre 1 year 2 data) as Comparator (below).

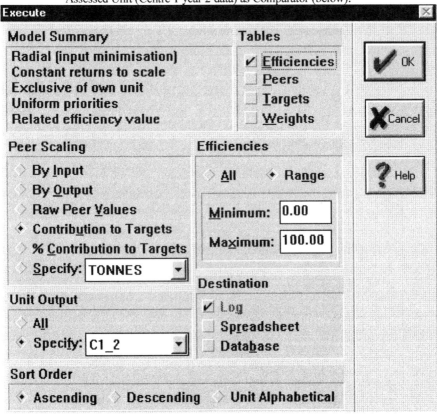

8.   CAPTURING THE IMPACT OF SCALE SIZE CHANGES ON PRODUCTIVITY

When the units being assessed operate a technology where efficient production is not characterised by constant returns to scale the change in the productivity of a unit may be impacted inter alia by changes in scale size. Figure 7.8 illustrates the case. It depicts a technology where Y is output, X is input and the maximum level of output attainable both in period t and t+1 is $Y = X^{0.5}$. There has been therefore no movement of the boundary over time as the expression applies across both time periods.

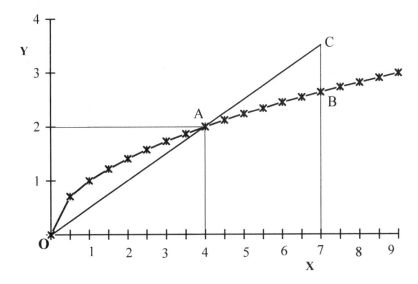

Figure 7.8. Measuring the Impact of Scale Size on Productivity

The technology in Figure 7.8 does not exhibit constant returns to scale. Let us assume now that we have a unit which in period t operated at point A (X = 4, Y = 2) and in period t+1 at B (X = 7, Y = $\sqrt{7}$). The productivity change as measured by the Malmquist Index as defined in Table 7.4 measures the <u>change in output per unit input</u> between t and t+1. Thus it yields productivity growth of ($\sqrt{7}$ / 7) / (2 / 4) = 0.76. This is less than 1 and we conclude that the operating unit has had productivity regress between period t and t+1. It produces less output per unit input in period t+1 than it did in period t. Note, however, that the operating unit's regress is entirely due to its change of scale of operation from X = 4 to X = 7 for the unit is operating on the technically efficient boundary of period t+1. If in period t+1 the operating unit is to yield the same output per unit input as it did in period t then it must operate at point C. That point is not feasible under the non-constant returns to scale technology operated by the unit. (Note that neither at A nor at B is the unit operating a most productive scale size but we simply focus here on comparing its productivity in years t and t+1.)

For units not operating under constant returns to scale as illustrated in Figure 7.8 it is possible to decompose their productivity change so that the impact of any scale size changes on productivity can be identified. The decomposition was first put forth by Färe et al (1994b). Denoting by $V\_EF_{Tt}^{Dt}$ the radial *pure* technical input efficiency of DMU j_0 (i.e. computed under VRS) and $SC\_EF_{Tt}^{Dt}$ its scale input efficiency we have, (see also Chapter 6, expression (6.2))

$$C\_EF_{Tt}^{Dt} = V\_EF_{Tt}^{Dt} \times SC\_EF_{Tt}^{Dt} \tag{7.5}$$

Using (7.5) within Table 7.4 we arrive at the decomposition of the Malmquist index MI_{j_0} of DMU j_0 shown in Table 7.5.

Table 7.5.   Capturing the Impact of Scale on the Malmquist Index of Productivity Change

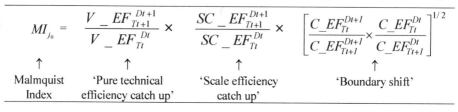

↑	↑	↑	↑
Malmquist Index	'Pure technical efficiency catch up'	'Scale efficiency catch up'	'Boundary shift'

The *pure technical efficiency catch up* term is interpreted in the same manner as the efficiency catch up term in Table 7.4. The only difference is that catch up is now measured relative to the efficient boundary corresponding to a variable rather than a constant returns to scale technology. Thus for example if the pure technical efficiency catch up term is over 1 it reflects efficiency progress in that the unit concerned is closer to the VRS boundary in period t+1 than it was to the respective boundary in period t. The interpretations of pure technical efficiency catch up values of 1 and less than 1 are analogous.

The *scale efficiency catch up* term reflects the extent to which DMU j_0 has become more scale efficient in between period t and t+1. **It is this term that captures the impact of any change in scale size of DMU j_0 on its productivity.** A value of 1 means scale efficiency is the same in periods t and t+1 and so DMU j_0 has had no impact to its productivity attributable to changes in its scale size. This does not necessarily mean that DMU j_0 has the same scale size in periods t and t+1. Rather, the impact of its scale size on its productivity is the same both in period t and in period t+1 ands so no change in this impact. When the scale efficiency change term is over 1 it means DMU j_0 is more scale efficient in period t+1 than it was in period t. This represents a positive gain to its productivity attributable solely to changes to its scale size between period t and t+1. Finally if the scale efficiency change term is under 1 it means DMU j_0 is less scale efficient in period t+1 than it was in period t and this in turn means that there is a negative impact on its productivity attributable to changes in its scale size.

Note that pure technical efficiency catch up and scale efficiency catch up are orientation *dependent*. This is intuitively correct. Scale size is measured on inputs when we adopt an output orientation and on outputs when we adopt an input orientation. Thus between two periods a DMU will have generally registered different change in scale size depending on the measure of scale size used. (This is not, of course, an arbitrary choice and depends on whether inputs or outputs are exogenous to the units being assessed.) The product of scale and pure technical efficiency catch up of DMU j_0 is (see (6.2) in Chapter 6) its technical input (or output) efficiency catch up (relative to a constant returns to scale boundary) and this is orientation *independent*.

The **boundary shift term** is the same in Tables 7.4 and 7.5. In both cases we measure the shift of the *constant* returns to scale boundary. The shift in the variable returns to scale boundary will generally differ from that of the CRS boundary.

7.9   THE COST MALMQUIST TYPE INDEX

The Malmquist index in Table 7.4 reflects input and output quantity data, whereas often it is not only the quantities of inputs and outputs but also the cost of inputs and the value of outputs that matter. Changes over time may impact both technology and factor prices which jointly may impact the scope for cost reduction or revenue expansion at a DMU. Thus where appropriate we would wish to reflect productivity changes in terms of the combined effect of input costs or output values relating to the two time periods over which productivity change is being measured. The *Cost Malmquist Type* index developed by Maniadakis and Thanassoulis (2000 and forthcoming) serves this purpose. The index can be computed either in the input orientation when input prices are available or in the output orientation when output values are available. We shall describe the index in its input orientation here. The adaptation to the output orientation is left to the reader.

Let DMU $j = 1...N$ use in period t inputs x_{ij}^t $(i = 1...m)$ to secure outputs y_{rj}^t $(r = 1...s)$. Let further the jth DMU face in period t input prices w_{ij}^t $(i = 1...m)$. We need to compute $C\_C_{Ttj_0}^{Dt}$, which is the least cost at which DMU j_0 could have produced its period t outputs y_{rj}^t $(r = 1...s)$ using the period t CRS technology (production possibility set) and input prices. $C\_C_{Ttj_0}^{Dt}$ is derived by solving model [M7.11].

$$C\_C_{Ttj_0}^{Dt} = \underset{x_i^t}{Min} \sum_{i=1}^{m} w_{ij_0}^t x_i^t \qquad\qquad [M7.11]$$

$$\text{Subject to: } \sum_{j=1}^{N} \lambda_j x_{ij}^t \le x_i^t \qquad i = 1...m$$

$$\sum_{j=1}^{N} \lambda_j y_{rj}^t \ge y_{rj_0}^t \qquad r = 1...s$$

$$\lambda_j \ge 0, j=1...N \ge 0, \ x_i^t \ge 0, \ \forall i.$$

Let us now denote $C\_IOE_{Ttj_0}^{Dt}$ the *input overall efficiency* of DMU j_0 in period t under the input prices it faces and CRS technology of period t. Then, (see also expression 4.3) we have:

$$C\_IOE_{Ttj_0}^{Dt} = \frac{C\_C_{Ttj_0}^{Dt}}{OC_{Ttj_0}^{Dt}} \qquad\qquad (7.6),$$

where $OC_{Ttj_0}^{Dt} = \sum_{i=1}^{m} w_{ij_0}^t x_{ij_0}^t$ is the observed cost at which DMU j_0 delivers its outputs in period t. The input overall efficiency of DMU j_0 measures the extent to which its observed cost $OC_{Ttj_0}^{Dt}$ can be reduced while the unit continues to produce its outputs of period t.

The *Cost Malmquist Type* (CMT$_i$) productivity index in the input orientation is

$$CMI_{j_0}^i = \left[\frac{C\_IOE_{Tt}^{Dt+1}}{C\_IOE_{Tt}^{Dt}} \times \frac{C\_IOE_{Tt+1}^{Dt+1}}{C\_IOE_{Tt+1}^{Dt}} \right]^{1/2} \qquad\qquad (7.7)$$

where $C\_IOE_{Ttj_0}^{Dt}$ is the input overall efficiency of DMU j_0 as defined in (7.6). Note that $C\_$ indicates that efficiencies are computed using CRS DEA models *irrespective* of the actual returns to scale characterising efficient production in the technology operated by the DMUs. This is so as to capture the impact of scale changes on productivity change as noted earlier with reference to the expression in (7.4). The cross period input overall efficiencies are computed by adapting suitably model [M7.11]. For example

$$C\_IOE_{Ttj_0}^{Dt+1} = \frac{C\_C_{Ttj_0}^{Dt+1}}{OC_{Ttj_0}^{Dt+1}} \text{ where } OC_{Ttj_0}^{Dt+1} = \sum_{i=1}^{m} w_{ij_0}^t x_{ij_0}^{t+1} \text{ and}$$

$$C\_C_{Ttj_0}^{Dt+1} = \underset{x_i^{t+1}}{\text{Min}} \sum_{i=1}^{m} w_{ij_0}^t x_i^{t+1} \qquad\qquad [M7.12]$$

Subject to:

$$\sum_{j=1}^{N} \lambda_j x_{ij}^t \leq x_i^{t+1} \qquad\qquad i = 1...m$$

$$\sum_{j=1}^{N} \lambda_j y_{rj}^t \geq y_{rj_0}^{t+1} \qquad\qquad r = 1...s$$

$$\lambda_j \geq 0, j=1...N \geq 0 , \ x_i^{t+1} \geq 0, \ \forall i.$$

The CMT index has a similar interpretation to the Malmquist index. It is a geometric mean of two indices. The first (i.e. the first ratio in the brackets in (7.7)) compares the two production points under evaluation by measuring their distance from the "cost technology" of period t and the second index does the same but with reference to the cost technology of period t+1. When $CMI_{j_0}^i > 1$ we have productivity progress in the sense that production is less costly in period t+1 for given output levels than in period t. Similarly, when $CMI_{j_0}^i < 1$ we have productivity regress in that production is more costly in period t+1 than in period t, for given output levels. Finally productivity is constant when $CMI_{j_0}^i = 1$.

It can be shown that when the input allocative efficiency of DMU j_0 is 1 we have $CMI_{j_0}^i = MI_{j_0}$, where MI_{j_0} is the Malmquist index as defined in Table 7.4. (See Maniadakis and Thanassoulis forthcoming). The index in (7.7) can be decomposed to capture allocative efficiency change, technical efficiency change and cost and technical boundary shift. However, these extensions of the CMT index are beyond the scope of this book. The interested reader can find them in Maniadakis and Thanassoulis (forthcoming).

10.   QUESTIONS

(Where the context permits the reader may employ the Warwick DEA Software accompanying this book to answer the questions below or to confirm answers derived by other means. See Chapter 10 for directions in using the limited version of the software accompanying this book.)

1.   Consider 11 DMUs which operate a constant returns to scale technology and produce the output levels in Table 7.6 per unit of the single input they use. The DMUs operate under two distinct policies as indicated. Ascertain the managerial and policy efficiency of each department. (NB. You may use graphical means and confirm your results using the *Warwick DEA Software.*)

Table 7.6.   Output Levels Per Unit of Input

	D1	D2	D3	D4	D5	D6	D7	D8	D9	D10	D11
Output 1	5.5	5.5	4	4	7	4	2	8	1	6.5	2.5
Output 2	8	5.5	7	3	6	9	8	3	8	1	9
Policy	1	2	1	1	2	1	2	2	2	1	1

2.   Table 7.7 shows 5 DMUs operating a constant returns to scale technology. The output levels shown are per unit of the single input they use and relate to two successive time periods of equal duration.

Table 7.7.   Output Levels Per Unit Input

	D1	D1	D2	D2	D3	D3	D4	D4	D5	D5
Output 1	5.5	4	5.5	7	4	8	1	2.5	4	2
Output 2	8	7	5.5	6	3	3	8	9	9	8
Period	1	2	1	2	1	2	2	1	1	2

a) Compute the productivity gain of DMU 4 between the two time periods.

b) To what extent has the boundary shifted at the output mixes operated by DMU 1 and to what extent has the DMU kept with any movement in that boundary?

3. Refer to the data sets labelled "Set 1" and "Set 2" below where DMUs are *rates* departments collecting local property taxes.

a) Assume that the DMUs in Set 1 operate a policy whereby certain self-contained functions such as salary payment and recruitment procedures are outsourced while departments in Set 2 operate a policy whereby all activities are carried out by own staff. Ascertain whether there is any evidence that outsourcing impacts efficiency.

b) Ignore part a) and assume instead that Sets 1 and 2 respectively contain data for the first and second quarter of the financial year for the same set of rates departments. Assume that department DMU1 in Set 1 corresponds to department DDMU1 in Set 2 and so on. Ascertain the productivity gain of each department between the two quarters and decompose it into that attributable to the department and that reflecting a shift of the efficient boundary between the two quarters.

Set 1	COST	HER-EDS	REBA-TES	SUM-MONS	VALUE
DMU1	6.340	8.168	16.613	8.264	5.047
DMU2	7.700	7.884	15.749	14.502	3.034
DMU3	5.990	5.666	27.546	5.243	3.410
DMU4	5.200	6.923	12.613	4.298	3.040
DMU5	6.360	7.352	23.510	5.744	4.207
DMU6	5.650	1.777	0.156	1.314	39.011
DMU7	21.600	15.107	70.958	54.216	10.809
DMU8	8.570	7.919	48.688	14.032	5.923
DMU9	6.010	7.066	36.304	5.445	2.936
DMU10	8.020	8.858	43.610	13.774	4.274
Set 2					
DDMU1	7.000	8.369	14.918	9.883	4.328
DDMU2	10.500	9.608	37.910	13.493	5.035
DDMU3	8.520	8.967	24.672	11.841	3.753
DDMU4	7.610	6.111	31.734	7.657	2.872
DDMU5	10.910	9.778	42.725	12.169	4.657
DDMU6	9.720	7.713	5.879	14.600	9.251
DDMU7	12.630	11.082	41.586	16.420	5.647
DDMU8	11.510	9.066	28.491	16.284	5.962
DDMU9	6.220	6.627	14.667	7.703	3.083
DDMU10	5.290	3.958	20.416	1.961	1.835

APPENDIX 7.1.

Input-output levels which would render the rates departments in Example 7.2 Pareto-efficient. The input-output levels have been computed under variable returns to scale in the input orientation, using *Warwick DEA Software*.

Urban Dept	COST	ACCOUNTS	REBATES	SUMMONS	VALUE
DEPT1	9.1	7.5	34.1	22	3.8
DEPT2	13.6	8.3	23.3	36	8.6
DEPT3	5.8	10.9	13.4	11.5	4.9
DEPT4	11.2	16.6	36.8	27.6	7.5
DEPT5	15.8	22.8	95.8	23.6	12.3
DEPT6	5.7	1.8	0.2	1.3	39
DEPT7	21.6	15.1	71	54.2	10.8
DEPT8	8.6	7.9	48.7	14	5.9
DEPT9	6	7.1	36.3	5.4	2.9
DEPT10	8	8.9	43.6	13.8	4.3
DEPT11	9.6	10.8	36.9	20.7	8.2
DEPT12	7.6	9.5	45.2	8.3	5.3
DEPT13	5.2	6.8	18.7	10.6	3.5
DEPT14	5.9	9	18.4	12.3	4.3
DEPT15	5.7	7.7	25.9	8.4	3.5
DEPT16	7.6	7.2	19.7	17.6	6.3
DEPT17	4.9	3.4	23.7	4.3	2.5
DEPT18	8.3	8.6	30.5	17.8	8
DEPT19	22	12.2	92	29.5	14.8
DEPT20	7.7	8.3	41.2	13.3	4.5
DEPT21	5.4	8.2	16.6	10.6	5
DEPT22	6.5	7.9	20.9	14.5	4.4
DEPT23	5.4	5.7	27.5	6.2	3.4
DEPT24	5.2	6.9	18.5	10.7	3.6
DEPT25	5.5	7.4	23.5	8.8	4.2
DEPT26	6.8	7.8	38.1	9.6	3.6
DEPT27	8.1	13.6	31.8	14.6	6.6
DEPT28	5.3	7.7	18	10.7	3.8
DEPT29	10.8	15.3	55.4	16.4	12.5
DEPT30	5.4	8.4	16.5	10.9	4.3
DEPT31	7.6	9.6	37.9	13.5	5

Rural Dept	COST	ACCOUNTS	REBATES	SUMMONS	VALUE
DEPT32	10.9	10.6	37	14.2	4.8
DEPT33	8.5	9	24.7	11.8	3.8
DEPT34	7.6	6.1	31.7	7.7	2.9
DEPT35	10.9	9.8	42.7	12.2	4.7
DEPT36	9.7	7.7	5.9	14.6	9.3
DEPT37	12.6	11.1	41.6	16.4	5.6
DEPT38	11.5	9.1	28.5	16.3	6
DEPT39	6.2	6.6	14.7	7.7	3.1
DEPT40	5.3	4	20.4	2	1.8
DEPT41	8.5	6.7	31.7	8.6	4.8
DEPT42	13.5	4.8	26.5	20.9	4.2
DEPT43	12.6	6.7	30.3	9.1	19.4
DEPT44	8.1	8.1	9.7	8.5	7.5
DEPT45	9	7.6	19.5	10.7	8
DEPT46	12.1	11.3	28.5	12.5	6.7
DEPT47	8.2	8.7	23.5	11.3	3.7
DEPT48	11.1	10.3	23	13.7	6.5
DEPT49	11.8	12.2	14.3	10.1	5
DEPT50	12.3	10.4	37.4	16.4	5.7
DEPT51	50.3	32.3	150	45.1	19.6
DEPT52	10.8	9.5	27.1	14.9	6.2
DEPT53	10.9	9	22	14.7	8.3
DEPT54	5.6	3.7	12.2	5.4	2.8
DEPT55	9.3	8.2	13.3	13.6	7.1
DEPT56	6.6	6.1	19.5	6.9	3.3
DEPT57	6.2	6	17.1	6.3	3.1
DEPT58	7.9	7.2	16.3	8.7	6.6
DEPT59	7.3	7.8	19.5	9.7	3.4
DEPT60	8.6	7.7	12.4	12.2	6.4
DEPT61	7.1	6.3	13.9	8.5	5.2
DEPT62	6.3	6.6	14.1	7.6	3.5

Chapter 8

INCORPORATING VALUE JUDGEMENTS IN DEA ASSESSMENTS

8.1 INTRODUCTION

We saw in Chapter 4 that we can give the DEA efficiency measure equivalently a production or value interpretation. In value-based interpretations the DEA efficiency measure of a DMU is the ratio of the sum of its weighted outputs to the sum of its weighted inputs. The weights used are DMU-specific and they are chosen so as to maximise its efficiency rating, subject only to the restriction that they should be positive. The imputation of input-output values in this way has a number of practical advantages. One such advantage is that the user need not identify prior relative values for inputs and outputs, permitting instead such values to be determined by the model solved.

Unfortunately, the imputation of input-output values in this manner can prove problematic in contexts where the user does have certain value judgements which should be taken into account in the assessment. The value judgements could reflect known information about how the factors of production used by the DMUs behave and/or 'accepted' beliefs of the relative worth of inputs and/or outputs. In such cases the input-output values imputed by the model may not accord with the value judgements at issue.

This chapter examines ways in which the foregoing problem can be overcome and the difficulties, advantages and drawbacks of the relevant approaches available.

8.2 WHY WE MAY WANT TO INCORPORATE VALUE JUDGEMENTS IN DEA ASSESSMENTS

We shall use the term *Decision Maker* (DM) to refer to the individual or to group of individuals conducting an assessment of the comparative efficiencies of a set of DMUs. Some of the circumstances where we would wish to incorporate value judgements in a DEA assessment are the following.

Imputed values may not accord well with prior views on the marginal rates of substitution and/or transformation of the factors of production.

As we saw in Chapter 4 (section 4.4.5) ratios of optimal input-output weights yielded by a DEA model such as [M4.5] are the *imputed* marginal rates of substitution or transformation between the inputs and/or outputs. Some of the imputed marginal rates of substitution or transformation may turn out to be ill-defined or counter-intuitive because some input and/or output weight takes the infinitesimal value ε. Further, even when not ill-defined, the marginal rates of substitution or transformation may not accord with the DM's prior views of the production process modelled. For example in assessing school effectiveness an 'unrestricted' DEA model such as [M4.5] could impute a higher value to the attainment of a lower grade by a pupil than to the attainment of a higher grade!

Certain inputs and outputs may have a special interdependence within the production process modelled.

For example Thanassoulis et al. (1995) assessing the efficiency of hospital perinatal care units in the UK wished to relate within the assessment the number of survivals of babies at risk to the number of babies at risk. They used "babies at risk" as an input and "survivals of babies at risk" as an output and required both to have the same weight. While the requirement that an input and output weight be equal in this manner is subjective, the approach permits the importance attached by the model to the ratio to be varied, but not the individual components of the ratio. A similar approach was adopted by Thanassoulis (1995) where in an assessment of police forces the weights on number of violent crimes cleared (an output) was related to the weight on the number of violent crimes reported (an input).

We may wish to arrive at some notion of 'overall' efficiency.

We saw in Chapter 4 that overall efficiency reflects both technical and allocative efficiency. Estimation of overall efficiency requires information on the prices of inputs or the worth of outputs. Such information is not always available, especially in not-for-profit contexts. Value judgements can be used as a proxy for such information. For example we can define a range of prices for inputs or worth of outputs within the context of a value-based DEA model to assess something akin to overall efficiency.

We may wish to discriminate between Pareto-efficient units.

The basic DEA model does not provide a direct means to discriminate between Pareto-efficient units. Incorporation of value judgements by means of restrictions on the relative worth of inputs or outputs can aid the discrimination between Pareto-efficient units. For example Thompson et al. (1986) in an attempt to site nuclear physics facilities in Texas found a lack of discrimination as five out of six alternative facilities were found relatively efficient by the basic DEA model. The discrimination of DEA was improved by defining ranges of acceptable weights, namely 'assurance regions', which were then used to select the preferred efficient site. (As we saw in Chapter 5, we may also use the frequency a peer is referred to by inefficient units and/or its contribution to targets as another way of discriminating between Pareto-efficient units.)

8.3   METHODS FOR INCORPORATING VALUE JUDGEMENTS IN DEA

Once the need is established to incorporate value judgements in a DEA assessment there are two broad types of method from which a choice can be made:
 ♦   Weights in the DEA model solved can be restricted;
 ♦   The comparative set of units can be altered in some way.

By far most of the available methods fall into the first category, that is restricting the DEA weights to reflect prior value judgements. We shall refer to this as the '*weights restrictions*' (WR) approach to incorporating value judgements in DEA assessments. The second category of methods above, that of manipulating the set of comparative units, is less well explored at the time of writing but does offer some potential either in combination with weights restrictions or as a stand-alone approach. Both approaches are introduced in this chapter.

8.4   USING   WEIGHTS   RESTRICTIONS   TO INCORPORATE VALUE JUDGEMENTS IN DEA

We will restrict our attention to the use of WR when the units being assessed operate a constant returns to scale technology. Weights restrictions under variable returns to scale have not been fully explored in the literature. One important issue is that in a variable returns to scale technology marginal rates of substitution are scale-dependent in principle and this complicates the framing of weights restrictions.

Weights restrictions may be applied to the DEA weights or to the product of the DEA weight with the respective input or output level, referred to as *virtual input* or *virtual output* (see section 4.5).

8.4.1 Restrictions Applied to DEA Weights

Let us assume that we have N DMUs ($j = 1...N$), each consuming varying amounts, x_{ij}, of m different inputs ($i = 1...N$) to produce varying quantities, y_{rj}, of s different outputs ($r = 1...N$). We assume that these quantities are strictly positive so that $x_{ij} > 0$ and $y_{rj} > 0$ $\forall i, r, j$. Model [M8.1] shows a comprehensive range of weights restrictions which can be used to incorporate value judgements in assessing the efficiency of DMU j_0.

$$\text{Max} \qquad \sum_{r=1}^{s} u_r y_{rj_0} \qquad\qquad \text{[M8.1]}$$

Subject to:

$$\sum_{i=1}^{m} v_i x_{ij_0} = C$$

$$\sum_{r=1}^{s} u_r y_{rj} - \sum_{i=1}^{m} v_i x_{ij} \leq 0 \qquad j = 1... j_0...N$$

$$\kappa_i v_i + \kappa_{i+1} v_{i+1} \leq v_{i+2} \qquad :r1$$

$$\alpha_i \leq \frac{v_i}{v_{i+1}} \leq \beta_i \qquad :r2$$

$$\mu_r u_r + \mu_{r+1} u_{r+1} \leq u_{r+2} \qquad :r3$$

$$\theta_r \leq \frac{u_r}{u_{r+1}} \leq \zeta_r \qquad :r4$$

$$\gamma_i v_i \geq u_r \qquad :r5$$

$$\delta_i \leq v_i \leq \tau_i \qquad :r6$$

$$\rho_r \leq u_r \leq \eta_r \qquad :r7$$

$$-v_i \leq -\varepsilon \qquad i = 1...m$$

$$-u_r \leq -\varepsilon \qquad r = 1...s$$

(ε is a non-Archimedean infinitesimal)

Model [M8.1] is derived from Model [M4.5] (Chapter 4) through the addition of the restrictions labelled r1 - r7 within the latter. Notation is the same in [M4.5] and [M8.1], u_r and v_i being the weights attaching respectively to the rth output and the ith input and they are the variables of the model. The Greek letters (κ_i, α_i, β_i, γ_i, δ_i, τ_i, ρ_r, η_r, θ_r, ζ_r, μ_r) are user-specified constants to reflect value judgements the DM wishes to incorporate in the assessment. They may relate to the perceived importance or worth of input and output factors. The restrictions in r1 - r7 are classified as follows:

a) Assurance regions of type I (ARI)

Restrictions of this type are illustrated by r1 - r4 in [M8.1]. Each restriction links either only input, or alternatively only output weights. The name Assurance Regions Type I is due to Thompson et al. (1990). Form r1 is similar to the type used in Thompson et al. (1986). Use of form r2 and r4 is more prevalent in practice, reflecting valid marginal rates of substitution as perceived by the DM. The upper or alternatively the lower of the bounds in r2 and r4 is often omitted.

b) Assurance regions type II (ARII)

This type of restriction is depicted by r5 within [M8.1]. Thompson et al. (1990) termed relationships between input and output weights 'Type II Assurance Regions' (ARII). ARII are typically used where some relationship between the output and input concerned is to be reflected in the DEA model (e.g. see Thanassoulis et al. 1995 and Thanassoulis 1995).

c) Absolute weights restrictions

These restrictions are illustrated by r6 and r7 in [M8.1] and are mainly introduced to prevent inputs or outputs from being over emphasised or ignored in the analysis. The meaning of the restriction is context dependent. For example an output weight may represent the marginal cost of a unit of the output concerned as elaborated in section 8.4 below. The levels of the bounds used in the restrictions are dependent on the normalisation constant, C in [M8.1] as C reflects an implicit scaling of the DEA weights. By the very nature of the DEA model in [M8.1] there is a strong interdependence between the bounds on different weights. For example, setting an upper bound on one input weight imposes implicitly a lower bound on the total virtual input of the remaining variables and this in turn has implications for the values the remaining input weights can take.

8.4.2 Restrictions Applied to Virtual Inputs and Outputs

We saw in Section 4.5 that the virtual inputs and outputs can be seen as normalised weights reflecting the extent to which the efficiency rating of a DMU is underscored by a given input or output variable. Rather than restricting the actual weight as outlined above some researchers, (e.g. Wong and Beasley 1990), suggest we could restrict virtual inputs and/or outputs. For example the proportion of the total virtual output of DMU j accounted for by output r can be restricted to lie in the range say $[\phi_r, \psi_r]$. The range is normally determined to reflect prior views on the relative 'importance' of the individual outputs. Thus the restrictions in the context of model [M8.1] would take the form

$$\phi_r \leq \frac{u_r y_{rj}}{\sum_{r=1}^{s} u_r y_{rj}} \leq \psi_r, \qquad r = 1 \ldots s \qquad \qquad :r8$$

A similar restriction can be set on the virtual inputs. Several approaches are suggested by Wong and Beasley (1990) for implementing restrictions on virtual values:

-   Add the restrictions only in respect of DMU j_0 being assessed leaving free the relative virtual values of the comparative DMUs;
-   Add the restrictions in respect of all the DMUs being compared. This is computationally expensive as the constraints added will be of the order of $2N(s + m)$;
-   Add the restrictions only in respect of the *average* DMU so that for example the restriction in respect of output r becomes

$$\phi_r \leq \frac{u_r \sum_{j=1}^{N} \frac{y_{rj}}{N}}{\sum_{r=1}^{s} u_r \left(\sum_{j=1}^{N} \frac{y_{rj}}{N} \right)} \leq \psi_r, \qquad r = 1 \ldots s \qquad \qquad :r9$$

where $\sum_{j=1}^{N} \frac{y_{rj}}{N}$ is the average level of the rth output across DMUs 1…N.

Restrictions on the virtual input-output weights represent indirect absolute bounds on the DEA weights of the type covered in *c)* in 8.3.1. Therefore as in the case of absolute bounds on the DEA weights, restrictions on the virtual inputs and outputs should be set in full recognition of the normalisation constant, C in [M8.1] as C reflects the implicit scaling of the virtual inputs and outputs. Restrictions on the virtual input/output weights have received relatively little attention in the DEA literature so far, but for an exploration of their impact in DEA assessments see Sarricco (1999).

8.5 SOME APPROACHES TO ESTIMATING THE PARAMETERS OF WEIGHTS RESTRICTIONS

A key difficulty in using weights restrictions is the estimation of the appropriate values for the parameters in the restrictions, (e.g. values for $\kappa_i, \alpha_i, \beta_i, \gamma_i, \delta_i, \tau_i, \rho_r, \eta_r, \theta_r, \zeta_r, \mu_r$ in [M8.1]). The parameters need to reflect the largely implicit DM value judgements in the DEA efficiency assessment. A number of methods have been put forward to aid the estimation of such parameters. However, no method is all-purpose and different approaches may be appropriate in different contexts. We outline here some of the better known approaches.

a)   Using unbounded DEA weights as reference levels
This approach is due to Roll et al. (1991) and Roll and Golany (1993). Initially an unbounded DEA model is run, a weights matrix compiled, and, if necessary, either the outlier weights or a certain percentage of the extreme weights are eliminated. Alternative optimal solutions generally exist, especially so for the Pareto-efficient DMUs. Hence, alternative weights matrices may exist and the choice of which matrix to use is for the user to make. The mean weight, U_r, V_i, for each factor is then calculated based on the selected weights matrix. A certain amount of allowable variation about each mean is subjectively determined, giving an upper and a lower bound for each factor weight.

b)   Using estimated average marginal rates of transformation as reference levels.
This approach is due to Dyson and Thanassoulis (1988) and it is applicable only to the case where DMUs use a single input to secure multiple outputs or alternatively secure a single output using multiple inputs.

Let us consider the case where the DMUs use a single resource, the jth DMU using x_j units to secure output levels y_{rj}, $(r = 1...s)$. Then the DEA weight on output r can be interpreted as the resource level assigned by the DEA model to each unit of output r. This can be seen more clearly if we set up model [M8.2], derived from the basic DEA model in [M4.5] for the single-input multiple-output case.

Max $\quad\displaystyle\sum_{r=1}^{s} u_r y_{rj_0}$ $\hspace{4cm}$ [M8.2]

Subject to:

$$v x_{j_o} = 1$$

$$\sum_{r=1}^{s} u_r y_{rj} - v x_j \le 0 \qquad j = 1...j_0 ...N$$

$$-u_r \le -\varepsilon \qquad r = 1...s$$

(ε is a non-Archimedean infinitesimal).

Notation in [M8.2] is as in [M4.5].

Let us set in [M8.2] $U_r = u_r x_{j_0}$, $\forall r$, eliminate the normalisation constraint by setting $v = 1/x_{j_0}$ and let us multiply the objective function and all constraints by $x_{j_0} > 0$. Model [M8.2] reduces to model [M8.3].

Max

$$\sum_{r=1}^{s} U_r y_{rj_0}$$

[M8.3]

Subject to:

$$\sum_{r=1}^{s} U_r y_{rj} \leq x_j \qquad j = 1 \dots j_0 \dots N$$

$$-U_r \leq -\varepsilon \qquad r = 1 \dots s$$

ε is a non-Archimedean infinitesimal.

At the optimal solution to model [M8.3] at least one constraint will be binding. Let the binding constraint relate to DMU j' so that $\sum_{r=1}^{s} U_r^* y_{rj'} = x_{j'}$. U_r^* is the optimal value of U_r in [M8.3].

We can interpret U_r^* within the expression $\sum_{r=1}^{s} U_r^* y_{rj'} = x_{j'}$ as the level of input the DEA model *allocates* per unit of output r.

This interpretation of the DEA weights enables us to identify reference levels for setting restrictions on the DEA weights. For example we could use regression analysis, regressing the input levels x_j on the output levels y_r ($r = 1 \dots s$) to estimate the average level of resource per unit of output r, using the input-output levels of the set of comparative DMUs ($j = 1 \dots N$). Lower bounds on the output weights can then be set with reference to the average resource or input levels per unit of output estimated.

For example let us assume that we estimate in the foregoing manner the regression equation

$$\bar{x} = \sum_{r=1}^{s} \phi_r y_r + \varsigma,$$

(8.1)

where ϕ_r is the partial regression coefficient of output r and ς is the regression constant. If $\varsigma = 0$ then the ϕ_r can be interpreted as the estimated resource units DMUs use on average per unit of output r. We can use the values of ϕ_r ($r = 1 \dots s$) as reference levels for setting lower or upper bounds on the DEA weights in [M8.3]. For example the user may decide to set lower bounds so that $U_r \geq 0.1 \phi_r$ ($r = 1 \dots s$). The argument would be that an efficient DMU cannot be so efficient as to use less than 10% of the resource level DMUs use on average per unit of the rth output. Alternatively, one may set upper bounds so that $U_r \leq 10\phi_r$ $(r = 1...s)$. The argument would be that however 'special' the operating practices of a DMU may be, they could not justify more than 10 times the level of resource other DMUs use on average per unit of the output concerned.

If ς is not zero in (8.1) but is not statistically significant then the model can be re-estimated forcing the regression constant to be zero. The resulting partial regression coefficients are then interpreted as above where $\varsigma = 0$. Finally if ς is statistically significantly different from zero then a variable returns to scale model may be more appropriate, though strictly speaking the equation in (8.1) related to average and not efficient operation.

Example 8.1: In Chapter 4, Example 4.3 we set up the value-based DEA model to assess the technical input efficiency of Centre 1.
a) Solve the model.

b) Let us assume that there is a prevailing view that in the context in which the Centres operate the procedure for delivering a package uses up at least the resource needed to travel 20 km.
   i.   Explain why it is necessary to modify the model solved in part a) to reflect this prevailing view.
   ii.   Modify the model solved in part a) to incorporate the prevailing view above.
   iii.   Solve the model as modified in (ii) and discuss the results obtained.
c) Reproduce the efficiency rating obtained in (biii) above using the *Warwick DEA Software*.

Solution: a) The model to assess the input efficiency of Centre 1 is [M8.4].

Max $\qquad Z_1 = 4.3u_D + 90u_P$ $\hfill$ [M8.4]
Subject to: $\quad 4.1v_L + 2.3v_C = 1$
Centre 1: $4.3u_D + 90u_P - 4.1v_L - 2.3v_C \leq 0$
Centre 2: $3.9u_D + 102u_P - 3.8v_L - 2.4v_C \leq 0$
Centre 3: $4.1u_D + 96u_P - 4.4v_L - 2v_C \leq 0$
Centre 4: $5.2u_D + 110u_P - 3.2v_L - 1.8v_C \leq 0$
Centre 5: $4.2u_D + 120u_P - 3.4v_L - 3.4v_C \leq 0$
$\qquad\qquad u_D, u_P, v_L, v_C \geq \varepsilon$, ε non-Archimedean infinitesimal.

Using $\varepsilon = 0.0001$ one solution to the above model is $u_D = 0.1505$, $u_P = 0.0001$, $v_L = 0.0001$ and $v_C = 0.43478$, yielding an efficiency of 0.6472.

b)

i.   The model solved should be modified because Centre 1 attains its maximum efficiency rating effectively ignoring the number of packages it delivers ($u_P = 0.0001$). This is intuitively unacceptable and contrary to the prevailing view that to deliver a package takes at least the resource needed to travel 20 km. So the model solved should be modified for a more realistic assessment of Centre 1.

ii.   Since the units in the model are thousands of km travelled and thousands of packages delivered then we can reflect the prevailing view above by adding to [M8.4] the constraint $20\,u_D \leq u_P$ This means one thousand delivered packages 'resourced' at u_P account for at least 20 thousand km travelled, 'resourced' at u_D per thousand km.

iii.   The solution to model [M8.4] after adding the constraint $20\,u_D \leq u_P$ is $u_D = 0.00035$, $u_P = 0.0071$, $v_L = 0$, $v_C = 0.4347$, efficiency rating $= 0.6403$. The imputed values for Centre 1 now equate each package delivered at about 21 km (i.e. 71 / 3.5) and this is within the prevailing view of the relative resource needed to deliver a package as opposed to travelling between successive delivery points. The efficiency rating of the centre could only go lower under the restricted weights. In the event the reduction to its efficiency rating is only marginal, down from 64.72% to 64.03%.

c) The input file for *Warwick DEA Software* is
    -LABOUR -CAPITAL +DISTANCE +PACKAGES
    C1 4.1 2.3 4.3 90
    C2 3.8 2.4 3.9 102
    C3 4.4 2 4.1 96
    C4 3.2 1.8 5.2 110
    C5 3.4 3.4 4.2 120
    END

We need to engage the *Warwick DEA Software* to solve model [M8.4], including the weights constraint $20\,u_D \leq u_P$. Model [M8.4] is radial input minimising (see Chapter 4 model [M4.5].) Thus in *Warwick DEA Software*

Options menu we select *Radial, Inputs*. To enter the constraint $20\ u_D \leq u_P$ we select the *Advanced* menu and enter the constraint as illustrated in Figure 8.1. (See Chapter 10 for more details on entering weights constraints in *Warwick DEA Software*.)

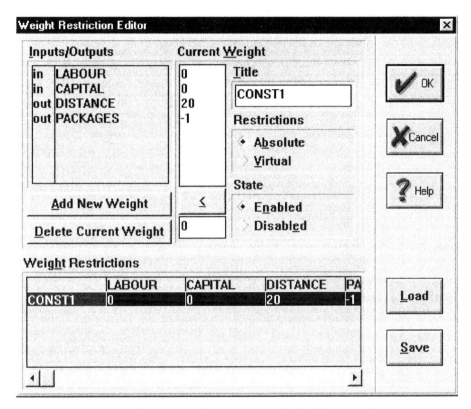

Figure 8.1. Imposing the Weights Restriction $20\ u_D - u_P \leq 0$

The efficiency rating *Warwick DEA Software* yields in respect of Centre 1 once the weights restriction in Figure 8.1 is imposed is 64.03%. This almost matches that computed in (biii) the difference being attributable to round-off errors.

8.6 INTERPRETING THE RESULTS OF DEA MODELS WITH WEIGHTS RESTRICTIONS

The interpretations of the results of DEA models do not in general carry over between '*unrestricted*' DEA models such as those in [M4.5] and [M6.8] and 'weights-restricted' models such as those discussed in this chapter. In particular, restricted and unrestricted models yield different targets and efficiency measures.

8.6.1 Effects of Weights Restrictions on the Interpretation of the Efficiency Measure Yielded By a DEA Model

As we have seen, when all DEA weights are unrestricted, the technical input efficiency rating of a DMU represents the maximum radial contraction to its input levels feasible without detriment to its output levels. (E.g. see the interpretation of k_0^* in model [M4.1] in Chapter 4.) Similarly, in the output orientation the efficiency rating of a DMU represents the maximum radial expansion to its output levels feasible without the need for additional resources. These interpretations of the efficiency rating do not carry over to the case where at least one of the DEA weights is restricted and the restriction is binding at the optimal solution to the DEA model solved.

The precise properties of the efficiency rating yielded by a weights-restricted DEA model would depend on the type of weights restrictions used. We consider here a simple weights restriction to illustrate the differences which arise between efficiency measures yielded by unrestricted and weights-restricted DEA models.

Without loss of generality consider the assessment of DMU j_0 by [M8.5].

Max $\qquad \sum_{r=1}^{s} u_r y_{rj_0}$ $\qquad\qquad$ [M8.5]

Subject to:

$$\sum_{i=1}^{m} v_i x_{ij_0} = C$$

$$\sum_{r=1}^{s} u_r y_{rj} - \sum_{i=1}^{m} v_i x_{ij} \le 0, \qquad j = 1...N$$

$$v_{i'} - f v_{i'+1} \ge 0 \qquad :r1$$

$$u_r, v_i \ge \varepsilon, \ \forall r \ and \ i.$$

Notation in [M8.5] is as in [M8.1]. We shall use the superscript * to denote optimal values. The efficiency rating yielded by [M8.5] is

$$\frac{\sum_{r=1}^{s} u_r^* y_{rj_0}}{\sum_{i=1}^{m} v_i^* x_{ij_0}} = \frac{\sum_{r=1}^{s} u_r^* y_{rj_0}}{C} = Z^*.$$ The following is now true:

If we contract radially the input levels of DMU j_0 by the factor Z^* without detriment to its output levels DMU j_0 will be rendered 100% efficient, but the resulting input-output levels may <u>not</u> lie within the PPS.

The proof of this can be found in Appendix 8.1. (A similar statement can be made in respect of expanding the output levels of DMU j_0. See Question 6 at the end of this chapter.)

Thus, when at least one weight restriction is binding in a DEA model we cannot use the DEA efficiency rating the model yields as a simple scaling constant to estimate expansions of output levels or contractions of input levels which are feasible in principle under efficient operation. Thus, the measure loses the type of operational meaning it had in the absence of weights restrictions. If the DEA efficiency measure under weights restrictions leads to a point outside the PPS in the manner suggested above, it would imply that the DMU concerned would need to alter its input or its output *mix* to attain 100% efficiency *within* the PPS, under the weights restrictions. (Mix refers here to the ratios the input and output levels are to each other.)

8.6.2 Effects of Weights Restrictions on DEA Targets

As we saw in Chapter 4 (expression 4.1), DEA models yield target input-output levels which would render a DMU Pareto-efficient. In the absence of weights restrictions such targets have the following two features:
- They preserve as far as possible its observed mix of input and output levels;
- They involve no deterioration to any observed input or output level.

These two features of targets can be lost when the targets are obtained from a DEA model which incorporates at least one binding weights restriction. For example Appendix 8.2 demonstrates that the targets yielded by DEA models to render a DMU Pareto-efficient under weights restrictions:
- can involve substantial changes to the DMU's current mix of input or output levels;
- can involve deterioration to the observed levels of certain of its inputs and outputs.

It should be noted that under weights restrictions these features are perfectly in line with intuition. As we now have prior views about the relative worth of inputs and outputs it is quite acceptable that for a DMU to attain maximum efficiency it may for example have to change the relative volumes of its activities (i.e. change its output mix). Further, the value judgements incorporated within the model may mean that by worsening the level of one output some other output can rise so as to more than compensate the loss of value due to the worse level on the former output.

8.6.3 Additional Effects of Absolute Weights Restrictions

When absolute weights restrictions are used in a DEA model, switching from an input to an output orientation can produce different relative efficiency scores, even under CRS. Hence the bounds need to be set in light of the model orientation used, which will flow out of the context of the DEA application and the degree of exogeneity of the input and the output variables. It should be recalled that unlike assurance regions, absolute weights restrictions may render a DEA model infeasible. Finally, Podinovski (1999) has shown that under certain conditions absolute weights restrictions will mean that the DEA model concerned may not lead to weights which give a DMU the best *ratio* of its own efficiency measure relative to that of any other DMU within the comparative set. In a sense this means the DEA weights under absolute weights restrictions may not enable a DMU to appear in the best possible light relative to other DMUs.

8.7   USING   UNOBSERVED   DMUS   TO INCORPORATE VALUE JUDGEMENTS IN DEA

As noted earlier, one major difficulty in using weights restrictions in DEA is the determination of appropriate formats and parameters for the restrictions so that they will capture the DM value judgements. This problem can be addressed in a different way by working on the *envelopment* rather than the *value-based* DEA model which could offer certain advantages.

DEA weights restrictions implicitly work on the envelopment DEA model in that they indirectly modify the PPS. (It is recalled that the PPS is the set containing all feasible input-output level correspondences pertaining to the production process operated by the DMUs.) Thanassoulis and Allen (1998) have demonstrated the equivalence between relative DEA weights restrictions (i.e. types ARI and ARII) and the incorporation of *Unobserved DMUs* (UDMUs) in DEA assessments. That is suitably constructed UDMUs can be added to the observed DMUs and then the efficiency rating obtained for each observed DMU within the aggregate set of observed and unobserved DMUs will be the same as if it had been assessed within the observed DMUs alone, under weights restrictions. Thus

UDMUs can in principle be used instead of weights restrictions to capture value judgements in DEA assessments.

A simple two-output one-input example, taken from Allen and Thanassoulis (1996), can be used to illustrate how adding UDMUs to those observed can implicitly capture certain DM value judgements. Consider the set of 11 DMUs shown in Table 8.1, each one yielding the output levels shown per unit of input.

Table 8.1.   Example Data Set

	D01	D02	D03	D04	D05	D06	D07	D08	D09	D10	D11
Output 1	5.5	5.5	4	4	7	4	2	8	1	6.5	2.5
Output 2	8	5.5	7	3	6	9	8	3	8	1	9

Figure 8.2 depicts the PPS generated by DMUs D01 - D11. The Pareto-efficient boundary consists of the facets D06D01, D01D05 and D05D08. Inefficient DMUs whose radial projections are on the inefficient frontier segments BD06 and CD08 would attain maximum efficiency rating by means of one output being given negligible (i.e. ε) weight. For example DMUs D09, D07 and D11 whose radial projections are on the inefficient part of the boundary BD06 would attain maximum DEA-efficiency rating by giving an ε-weight to output 1. (For the correspondence between inefficient parts of the PPS boundary and ε-weights in DEA see Lang et al. 1995).

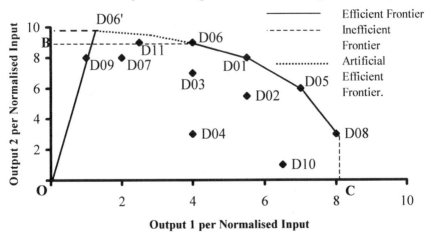

Figure 8.2. Extended Production Possibility Set

Assume now that the DM finds it unacceptable that any output should in effect be ignored by any DMU in the assessment, but has no clear view as to what might be a sensible lower bound on each weight. Note, however, that if the DM can trade-off between the output levels of D06 reducing the level of output 1 while raising the level of output 2 to compensate, a UDMU can be created say at D06′ in Figure 8.2. This would extend the Pareto-efficient part of the PPS boundary and DMUs such as D07, D09 and D11 will no longer be using an ε-weight on output 1 to attain maximum DEA efficiency rating.

Generalising this approach Allen and Thanassoulis (1996) and Allen (1997) develop a number of practical algorithms for reducing the DEA-inefficient boundary of the PPS and thereby reducing the number of ε-weights used in arriving at the efficiency rating of DMUs. This is equivalent to implicitly imposing lower bounds on DEA weights in line with DM judgements so as to avoid the use of ε-weights in arriving at the DEA efficiency rating of DMUs.

8.7.1 A Procedure for Improving Envelopment in DEA

The ideas illustrated in Figure 8.2 are developed in Allen and Thanassoulis (1996) into a general purpose procedure for '*improving envelopment*' in DEA.

A fully enveloped DMU is one which attains maximum efficiency without using ε-weights for any input or output.

The procedure addresses the case where the DMUs operate a constant returns to scale technology in a single input multi-output context. We sketch the procedure here. The related mathematical proofs are beyond the scope of this book but may be found in Allen and Thanassoulis (1996).

The procedure developed in Allen and Thanassoulis (1996) for improving envelopment in DEA it is as follows:

(i) Run an unrestricted assessment by DEA to identify the Pareto-efficient and non-enveloped DMUs. Stop if all DMUs are fully enveloped.

(ii) If any non-enveloped DMUs exist identify *Anchor* DMUs (ADMUs) from which to construct *Unobserved* DMUs (UDMUs).

(iii) In respect of each ADMU identify which output(s) to adjust in order to construct suitable UDMUs.

(iv) Using the outputs in (iii) and DM value judgements construct UDMUs.

(v) Re-assess the observed DMUs by DEA after adding the UDMUs constructed. The number of fully enveloped observed DMUs will generally increase.

The foregoing steps are implemented as follows:

- *Step (i): Assessing the Dmus*

Consider a set of DMUs (j = *1...N*), each using under CRS varying amounts of a single input to secure varying quantities of s different outputs. Let y_{rj} be the level of output r *per unit* of input secured by DMU j. Use an unrestricted DEA model (e.g. [M4.5]) to assess the DMUs. If all DEA-inefficient DMUs are properly enveloped in the sense that none accords any output a weight of ε then stop. Otherwise proceed to step (ii).

- *Step (ii): Identify anchor DMUs*

Anchor DMUs, (ADMUs) are a sub-set of the Pareto-efficient DMUs which, normally, but not always, delineate the Pareto-efficient from the inefficient part of the PPS boundary. (E.g. DMUs D06 and D08 in Figure 8.2 are ADMUs.) ADMUs are special because relatively minor (or local) adjustments to their input-output levels can reduce the DEA-inefficient part of the PPS and thereby increase the number of fully enveloped DMUs.

In the general case it is necessary to solve a linear programming model to ascertain whether a Pareto-efficient DMU is an ADMU. The model solved is a variant of that introduced by Andersen and Petersen (1993). Let the set JE consist of the Pareto-efficient DMUs identified in Step (i) and let JE_{j_0} be the set JE excluding DMU $j_0 \in JE$. In respect of each $j_0 \in JE$ solve the following envelopment model:

$$h'_{j_0} = \underset{\phi_0, \lambda_{,j}, G_r}{Min} \; \phi_0 - \varepsilon \sum_{r=1}^{s} G_r \qquad [M8.6]$$

Subject to:

$$\sum_{j \in JE_{j_0}} \lambda_j = \phi_0$$

$$\sum_{j \in JE_{j_0}} \lambda_j y_{rj} - G_r - y_{rj_0} = 0 \qquad r = 1...s$$

$$G_r, \lambda_j \geq 0 \qquad r = 1...s \quad j \in JE_{j_0}.$$

Model [M8.6] is as [M4.1] (Chapter 4) but using only the Pareto-efficient DMUs identified in Step (i), excluding DMU j_0. (The input is normalised to one unit.)

DMU j_0 is an ADMU if the optimal value of ϕ_0 in [M8.6] is $\phi'_0 > 1$ and there is at least one positive slack variable G_r.

Fuller details on the rationale underlying ADMUs can be found in Allen and Thanassoulis (1996).

• *Step (iii): Determining which output levels of the ADMUs to adjust*
Let K_{j_0} be the set of outputs for which the instance of model [M8.6] corresponding to ADMU j_0 yields zero slack. For each $k \in K_{j_0}$ solve model [M8.7]. Notation in [M8.7] is as in [M8.6], ϕ_0^* being the optimal value of $\square_0$ in [M8.6]. If there exists an optimal solution to the instance of [M8.6] corresponding to ADMU j_0 in which output k has a positive slack value, then G_k would be positive at the optimal solution to [M8.7].

Max G_k [M8.7]
Subject to:

$$\sum_{j \in JE_{j_0}} \lambda_j = \phi_0^*$$

$$\sum_{j \in JE_{j_0}} \lambda_j y_{rj} - G_r - y_{rj_0} = 0 \quad r = 1...s$$

$$G_r, \lambda_j \geq 0 \qquad r = 1...s \quad j \in JE_{j_0}.$$

UDMUs are constructed by setting to a 'lowest permissible' level each output of ADMU j_0 in turn for which either model [M8.6] or model [M8.7] yields a positive slack value at its optimal solution.

In the general case the minimum permissible level my_r of output r per unit of input will reflect what the DM deems is feasible under both technological and policy constraints. In many instances it will be the case that $my_r = 0$. However, in some practical contexts zero output levels are impossible or simply not acceptable.

• *Step (iv): Constructing suitable UDMUs*
For each ADMU and output level reduced in Step (iii) a UDMU is constructed by suitably compensating for that output's reduction. This is done using implicit, local trade offs between output levels, given by the DM. Two possibilities exist. The DM has:

♦   no preferences over the <u>relative</u> changes of the output levels;
♦   preferences over the <u>relative</u> changes of the output levels.

Where the DM has no preferences over the relative changes of the output levels the DM can be merely asked to specify the reduction in the level of the input of the ADMU which will compensate for the reduction to the minimum level specified for one of the outputs, so *that the ADMU and the UDMU created will be deemed* (subjectively by the DM) *to be equally efficient.*

Where the DM does have preferences over the relative changes in output levels the DM is asked for the percentage p_k by which the level of the kth output needs to rise so that in the DM's view the raised levels of the s-1 outputs *jointly* compensate for the reduction to the minimum level specified for the rth output.

- *Step (v): Reassess the observed DMUs including within the comparative set the UDMUs constructed*

Once the UDMUs in step (iv) have been created, the observed DMUs are assessed using an unrestricted DEA model such as [M4.5], but within the combined comparative set consisting of the observed DMUs and the UDMUs created. The number of observed DMUs which are fully enveloped would normally be larger than in the absence of the UDMUs. For the proof of this see Appendix 2 in Allen and Thanassoulis (1996).

The increase in the number of fully enveloped DMUs obtained will depend largely on how successful the DM has been in giving trade-offs between the output levels of ADMUs which lead to Pareto-efficient UDMUs. In addition, it will also depend on the presence of any DMUs which are not properly enveloped by virtue of the fact that they make reference to *non-full dimensional efficient facets* (NFDEFs). (See Olesen and Petersen 1996).) The larger the number of DMUs making such reference the less likely is the procedure to lead to their envelopment as it can fail to detect ADMUs which can be used to construct UDMUs which will create *full dimensional efficient facets* FDEFs. Finally, if the minimum level my_r used in creating UDMUs, is not as low as the lowest level of output r per unit input found at some non-enveloped DMU, then such a DMU will always fail to be enveloped properly.

8.7.2 Advantages and Drawbacks UDMUs in Incorporating Value Judgements in DEA.

Some advantages of using UDMUs to capture value judgements are:

- The trade offs the DM is asked to specify when constructing a UDMU are local.

The use of preferences information local to some DMU rather than global may be easier for some decision makers. Moreover, global preferences in linear form as used in weights restrictions are too restrictive and may not capture marginal rates of substitution or transformation which vary within the PPS. This is likely to be especially true in VRS technologies. Local trade-offs between input-output levels can be specified implicitly by means of comparing similar DMUs which some DMs find easier than specifying explicit numerical trade-offs.

♦   **The use of UDMUS makes the modification to the PPS through the introduction of value judgements explicit.**
This offers the advantage that the feasibility of the modified PPS can be assessed by the DM. In contrast, under weights restrictions, the modification of the PPS is implicit and therefore its feasibility cannot be ascertained by the DM. The DM can construct the UDMUs taking explicit account of information on any technological or policy limitations to production.

♦   **The explicit modification of the PPS allows a more realistic set of targets to be set to render Pareto-efficient previously inefficient DMUs.**
As noted earlier, under weights restrictions the targets estimated for an inefficient DMU need not maintain the mix of its inputs or outputs and may mean certain input levels need to rise or output levels need to fall. The introduction of UDMUs retains the radial nature of targets albeit at the expense that such targets may lie on a part of the efficient boundary which makes reference to UDMUs. Such a boundary contains points which are *judgementally* feasible but not feasible by the normal definition of the PPS. Radial targets of this nature offer an alternative route to efficiency for an inefficient DMU. That is the DMU can attain efficiency by improving the absolute levels of its input-output bundle but keeping the mix constant, rather than by needing to alter the mix and the absolute levels of its input-output bundle.

Drawbacks of using UDMUs to capture value judgement are:

♦   **The approach may prove very demanding of DM time.**
Depending on the size of the problem being solved, both in terms of number of DMUs and input-output variables, there could be the need to construct many UDMUs. The construction of each UDMU requires the involvement of the DM, who may thus find the process elaborate.

♦   **The DM may find it difficult to provide the information required.**
Although the information required by the method is local and in the form of comparisons of DMUs, the DM could still find it hard to supply it.

8.8.   QUESTIONS

(Where the context permits the reader may employ the Warwick DEA Software accompanying this book to answer the questions below. See Chapter 10 for a limited User Guide.)

1.   A family business owns five restaurants and wishes to compare them for their relative effectiveness in attracting a share of their local restaurant meals market. The family decided to use DEA and collected the following data for the latest full quarter:

Table 8.2.   Latest Full Quarter of the Year

Restaurant	Local restaurant meals market ($00,000)	Floor space (m²)	Turnover ($000)
R1	3	100	80
R2	7	150	145
R3	5	140	200
R4	4	170	60
R5	1	80	50

The size of the local market represents the total turnover of restaurants in the locality of each restaurant owned by the family.

a)   Write down a CRS unrestricted weights-based DEA model which could be solved to assess the relative efficiency of restaurant R1.

b)   If the weights-based DEA model in a) is solved it is found that its efficient peers are restaurants R3 and R5. Without solving the model derive the DEA efficiency rating of restaurant R1.

c)   Assume that $100,000 in local restaurants meals market is worth no more than 100 m² of floor space in terms of enabling a restaurant to attract customers. Explain whether or not the efficiency derived in b) remains valid and if not derive the new efficiency rating for R1, solving the appropriate DEA model.

d)   Suggest in brief how the comparison of the restaurants on their ability to attract a share of their local market can be made fairer by collecting data on additional variables.

2.   The Circulation Manager of a newspaper has decided to use DEA to compare six of the retailers it supplies with papers to identify those who are more effective in minimising wasted unsold copies. The manager believes that the lower the sales volume and sales volatility of a retailer the lower their wasted copies should be. He uses the range of sales volumes recorded by each retailer as a measure of sales volatility. This range is the difference between the highest and the lowest volume of papers sold by the retailer in the latest full calendar month. The manager has collected the following data in respect of the last full calendar month.

Table 8.3. Newspaper Circulation Data

Retailer	Wasted copies	Mean daily sales	Range
RT1	8	85	20
RT2	5	45	12
RT3	7	65	8
RT4	6	78	10
RT5	3	45	4
RT6	4	56	18

a) Set up and solve a CRS unrestricted weights-based DEA model to assess the efficiency of retailer RT1.

b) Determine a set of input-output levels which would render RT1 Pareto-efficient.

d) In the model formulated in part (a) restrict the weight on sales to be no more than twice that on range. Solve the model and comment on the efficiency obtained now compared to that in part (a).

3.   Reconsider sales representatives 3 of question 5, Chapter 5.
Identify the trade-offs (marginal rate of substitution) between value of sales and households signed up which maximise the input efficiency of representative 3. Comment on the intuitive appeal of the trade-offs.

a) Determine the DEA-efficiency rating of sales representative 3 when one household signed up cannot be worth more than $10,000 or less than $1000 in sales secured. Comment on the result obtained now relative to the result in (a).

b) Identify sets of targets which would render sales representative 3 efficient respectively under the unrestricted trade-offs in part (a) and under relative worth between sales and households signed up as specified in (b). Comment on any differences between the two sets of targets.

4.   Reconsider the five residential homes of question 4, Chapter 5, revising the data for Home B so that the full set of data is now as in Table 8.4.

Table 8.4. Revised Data for Residential Care Homes

Home	A	B	C	D	E
Sdays (000)	10	11	14	21	10
Ldays (000)	9	11	17	2	5
Running costs ($000)	100	85	150	90	78

Assuming there are constant returns to scale between running costs and number of short and long stay days:

a) Identify both graphically and by solving the appropriate DEA models care homes which are not fully enveloped;

b) Identify any anchor care homes and the outputs for each anchor care home that would in principle need to be adjusted in order to improve envelopment. Do this both graphically and by solving the appropriate DEA models;

c) Attempt to improve envelopment by specifying appropriate unobserved DMUs. Comment on your results.

5.   With reference to the DMUs in Appendix 5.2 suggest a way of estimating bounds for DEA weights. Suggest some actual weights bounds and use them to assess the relative efficiencies of two DMUs of your choice. Comment on the results obtained in terms of the efficiency ratings and the targets for efficiency estimated.

6.   Let Z^* be the efficiency rating yielded by model [M8.5]. Prove that if we expand radially the output levels of DMU j_0 by the factor $1 / Z^*$ the DMU will be rendered 100% efficient, but the resulting input-output levels may not lie within the PPS.

APPENDIX 8.1: SOME FEATURES OF THE DEA EFFICIENCY MEASURE UNDER WEIGHTS RESTRICTIONS

Let us assess DMU j_0 using model [M8.5] and let the resulting efficiency rating be $\dfrac{\sum\limits_{r=1}^{s} u_r^* y_{rj_0}}{\sum\limits_{i=1}^{m} v_i^* x_{ij_0}} = Z^*$. The superscript * denotes optimal values. Then if we contract radially the input levels of DMU j_0 by the factor Z^* without detriment to its output levels DMU j_0 will be rendered 100% efficient, but the resulting input-output levels may <u>not</u> lie within the PPS.

Proof
Consider keeping the output levels of DMU j_0 constant while its ith input level becomes $x'_{ij_0} = Z^* x_{ij_0}$, i=1...m. Since $\sum\limits_{r=1}^{s} v_i^* x_{ij_0} = C$ we will have $\sum\limits_{r=1}^{s} v_i^* x'_{ij_0} = Z^* C$. The optimal solution to [M8.5] is feasible under the revised input levels of DMU j_0 because its corresponding constraint in [M8.5] becomes $\sum\limits_{r=1}^{s} u_r^* y_{rj_0} - \sum\limits_{r=1}^{s} v_i^* x'_{ij_0} = Z^* C - CZ^* = 0$. Further, with the revised input levels DMU j_0 has efficiency rating $\dfrac{\sum\limits_{r=1}^{s} u_r^* y_{rj_0}}{\sum\limits_{i=1}^{m} v_i^* x'_{ij_0}} = \dfrac{CZ^*}{CZ^*} = 1 \cdot$

Thus, DMU j_0 becomes 100% efficient when its input levels are scaled down as above.

We can use model [MA8.1.1] which is the dual to model [M8.5] to show that the radially contracted input levels of DMU j_0 as above may not lead to a point within the PPS. By virtue of duality the optimal objective function values of [MA8.1.1] and [M8.5] are equal and so since we would use ε such that $\varepsilon \ll C$ we would have $\sum_{r=1}^{s} u_r^* y_{rj_0} = CZ^* \approx Ch^*$. (The approximation is closer the smaller the value of ε.) Thus the input-oriented radial efficiency measure h^* yielded by [MA8.1.1] is approximately the same as the efficiency measure $\dfrac{\sum_{r=1}^{s} u_r^* y_{rj_0}}{\sum_{i=1}^{m} v_i^* x_{ij_0}} = Z^*$ yielded by [M8.5].

Min $\qquad Ch - \varepsilon\left(\sum_{r=1}^{s} S_r^+ + \sum_{i=1}^{m} S_i^- \right)$ $\qquad$ [MA8.1.1]

Subject to:

$$\sum_{j=1}^{N} y_{rj} \lambda_j - S_r^+ = y_{rj_0} \qquad r = 1 \dots s$$

$$h x_{ij_0} - \sum_{j=1}^{N} x_{ij} \lambda_j - S_i^- = 0 \quad i = 1,\dots,i'-1, i'+2,\dots,m$$

$$h x_{i'j_0} - \sum_{j=1}^{N} x_{i'j} \lambda_j - q - S_{i'}^- = 0$$

$$h x_{i'+1j_0} - \sum_{j=1}^{N} x_{i'+1j} \lambda_j + fq - S_{i'+1}^- = 0$$

h free, q, $\lambda_j \geq 0, j = 1 \dots N$.

The optimal solution to model [MA8.1.1] yields in respect of DMU j_0 targets input-output levels ($y_r^t, x_i^t \ i \neq i', i \neq i'+1, x_{i'}^t, x_{i'+1}^t$) where

$$y_r^t = \sum_{j=1}^{N} y_{rj} \lambda_j^* = S_r^{+*} + y_{rj_0} \qquad r = 1 \dots s \qquad\qquad (A8.1.1)$$

$$x_i^t = \sum_{j=1}^{N} x_{ij} \lambda_j^* = h^* x_{ij_0} - S_i^{-*} \quad i = 1,\dots,i'-1, i'+2,\dots,m$$

$$x_{i'}^t = \sum_{j=1}^{N} x_{i'j} \lambda_j^* = h^* x_{i'j_0} - S_{i'}^{-*} - q^*$$

$$x_{i'+1}^t = \sum_{j=1}^{N} x_{i'+1j} \lambda_j^* = h^* x_{i'+1j_0} - S_{i'+1}^{-*} + fq^*.$$

The targets in $(y_r^t, x_i^t \, i \neq i', i \neq i' + 1, x_{i'}^t, x_{i'+1}^t)$ render DMU j_0 Pareto-efficient.

To see this note that no increase to any output level or decrease in any input level in $(y_r^t, x_i^t \, i \neq i', i \neq i' + 1, x_{i'}^t, x_{i'+1}^t)$ is feasible without a compensatory decrease in some output level and/or rise in some input level for it would contradict the optimality of the slack values S_r^+, S_i^- $\forall i$ and r in [MA8.1.1].

Note now that when the weights restriction r1 in [M8.5] is binding, q^* will normally take a positive value. This could imply that we have $\sum_{j=1}^{N} x_{i'+1j} \lambda_j^* > h^* x_{i'+1j_0}$. When this happens $h^* x_{i'+1j_0}$ will not be a feasible level for input i'+1 in (A8.1.1) as it reduces the Pareto-efficient level $\sum_{j=1}^{N} x_{i'+1j} \lambda_j^*$ for that input without detriment to another input or output level.

Thus if we scale the observed input levels of DMU j_0 down so that $x_{ij_0}' = Z^* x_{ij_0} \cong h^* x_{ij_0}$, i=1...m, without detriment to its outputs we may not arrive at a point within the PPS.

APPENDIX 8.2: SOME FEATURES OF TARGETS YIELDED BY DEA MODELS UNDER WEIGHTS RESTRICTIONS

Let us assess DMU j_0 using model [M8.5]. The resulting targets that will render DMU j_0 Pareto-efficient are as defined in (A8.1.1), Appendix 8.1. The following are true:

♦ The preservation of the observed mix of input-output levels of DMU j_0 is not given pre-emptive priority within the targets in (A8.1.1).

The input-output mix of DMU j_0 within the targets in (A8.1.1) would be fully preserved if the slacks S_r^+, S_i^- $\forall i$ and r and q in (A8.1.1) were zero. The preservation of this mix is pursued within [MA8.1.1] only to the extent that the minimisation of h has pre-emptive priority over the maximisation of the sum of the slack values. However, the minimisation of h in [MA8.1.1] has no pre-emptive priority over the value q takes. The larger the value of q at the optimal solution to [MA8.1.1] the more dissimilar the input-output mix of DMU j_0 and of its targets in (A8.1.1).

♦ In respect of DMU j_0 a target input level may be higher (i.e. worse) than the corresponding observed input level.

This can be readily seen in the target level for input i'+1 in (A8.1.1). Clearly to the extent that fq^* could be positive we could have $x_{i'+1}^t > x_{i'+1_{j_0}}$.

Chapter 9

EXTENSIONS TO BASIC DEA MODELS

9.1 INTRODUCTION

The *basic* DEA models (e.g. models [M4.1] and [M6.1] for input efficiency under CRS and VRS respectively) have been extended in a number of ways so that they may become appropriate for assessing efficiency in more specialised contexts. The literature on modifications of the basic DEA models is extensive and growing. We shall cover in this chapter three of the earliest extensions to basic DEA models which address the following situations:
- Some of the input and/or output variables are exogenously fixed;
- The DM does not have uniform preferences over improvements to input-output levels which would render a DMU Pareto-efficient;
- Some of the input and/or output variables can only take categorical values.

In these contexts the basic DEA models are inadequate as we shall see and the modifications developed capture more accurately the decision making context.

9.2 ASSESSING EFFICIENCY UNDER EXOGENOUSLY FIXED INPUT-OUTPUT VARIABLES

Basic DEA models (e.g. [M4.1] and [M6.1]) measure the efficiency of a DMU in terms of the maximum radial contraction to its input levels or expansion to its output levels feasible under efficient operation. Such measures are not suitable in contexts where at least one of the variables to be radially contracted or expanded is *exogenously* fixed, in the sense that it is not controllable by the DMU.

Banker and Morey (1986a) extended basic DEA models so that they can be used for assessing efficiency under exogenously fixed input-output variables. Figure 9.1 helps to illustrate their approach.

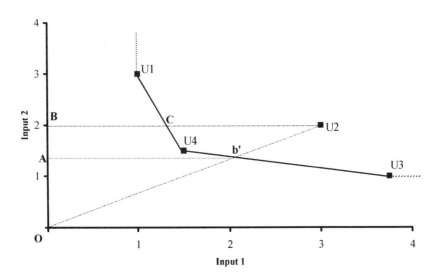

Figure 9.1.   Measuring Input Efficiency when Input 2 is Exogenously Fixed

Figure 9.1 depicts a set of DMUs operating a CRS technology using two inputs to secure one output. Output has been standardised to one unit. Let us assume that input 1 is controllable by the DMU but input 2 is exogenously fixed. The basic input-oriented DEA model [M4.1] uses b' as an efficient referent point for DMU U2 and so the efficiency measure it yields is $\dfrac{Ob'}{OU2}$.

This measure, however, reflects the uniform contraction to its input levels per unit of output, where the implied reduction to the level of input 2 is AB. This reduction to the level of input 2 is not, however, feasible because input 2 is exogenously fixed. The extension to the basic DEA model put forth by Banker and Morey (1986a) assesses the efficiency of DMU U2 with reference to the maximum feasible contraction to input 1, which is not exogenously fixed, controlling for the level of input 2, which is exogenously fixed. Thus the Banker and Morey (1986a) extended DEA model would measure the efficiency of DMU U2 in Figure 9.1 with reference to point C. The resulting efficiency measure will be $\dfrac{BC}{BU2}$.

Models [M9.1] and [M9.2] show respectively the modifications to the basic input oriented VRS and CRS DEA models developed by Banker and Morey (1986a) for dealing with exogenously fixed input-output variables in the general case. The corresponding output oriented models can be deduced by analogy. (See the Question 5 in Section 9.5.)

Model [M9.1] relates to DMUs ($j = 1...N$) using m inputs to secure s outputs. We denote x_{ij}, and y_{rj} the level of the ith input and rth output respectively at DMU j. We partition the inputs into two subsets I_d and I_{nd} consisting respectively of *non-discretionary* (i.e. exogenously fixed) and *discretionary* (i.e. under managerial control) inputs. If the DMUs operate under VRS then the *pure* technical input efficiency of DMU j_0 is h_0^*, where h_0^* is the optimal value of h_0 in

Min $\qquad h_0 - \varepsilon \left[\sum_{i \in I_d} S_i^- + \sum_{r=1}^{s} S_r^+ \right]$ $\qquad$ [M9.1]

Subject to:

$$\sum_{j=1}^{N} \lambda_j x_{ij} = h_0 x_{ij_0} - S_i^- \qquad i \in I_d \quad \text{(under managerial control)}$$

$$\sum_{j=1}^{N} \lambda_j x_{ij} = x_{ij_0} - S_i^- \qquad i \in I_{nd} \quad \text{(exogenously fixed)}$$

$$\sum_{j=1}^{N} \lambda_j y_{rj} = S_r^+ + y_{rj_0} \qquad r = 1...s$$

$$\sum_{j=1}^{N} \lambda_j = 1$$

$$\lambda_j \geq 0, j = 1...N, S_i^-, S_r^+ \geq 0 \,\forall\, i \, and \, r, h_0 \text{ free.}$$

ε is a non-Archimedean infinitesimal.

Model [M9.1] differs from the basic DEA model [M6.1] for assessing pure technical input efficiency under VRS in two respects. Firstly, h_0 measures the radial contraction feasible *only* in respect of the discretionary rather than all inputs. Secondly, the slacks of exogenously fixed inputs are not reflected within the summation term in the objective function. Thus

the efficiency measure arrived at by means of [M9.1] reflects the extent to which DMU j_0 can contract radially only those of its inputs which are NOT exogenously fixed.

Note that model [M9.1] yields target input-output levels which would render DMU j_0 Pareto-efficient *only* in respect of the outputs and the inputs which are not exogenously fixed. The levels of one or more of the exogenously fixed inputs may be capable of becoming lower without detriment to the DMU's controllable inputs or its outputs. Potential reductions to exogenously fixed inputs are not captured by model [M9.1] because the summation term within its objective function does not involve the corresponding slack variables.

Assume now that the DMUs relating to model [M9.1] above operate a CRS technology. Then to assess the technical input efficiency of DMU j_0 model [M9.2] is used.

$$\text{Min} \qquad h_0 - \varepsilon \left[\sum_{i \in I_d} S_i^- + \sum_{r=1}^{s} S_r^+ \right] \qquad\qquad \text{[M9.2]}$$

Subject to:

$$\sum_{j=1}^{N} \lambda_j x_{ij} = h_0 x_{ij_0} - S_i^- \qquad i \in I_d \ \text{(under managerial control)}$$

$$\sum_{j=1}^{N} \lambda_j x_{ij} = \sum_{j=1}^{N} \lambda_j x_{ij_0} - S_i^- \qquad i \in I_{nd} \ \text{(exogenously fixed)}$$

$$\sum_{j=1}^{N} \lambda_j y_{rj} = S_r^+ + y_{rj_0} \qquad r = 1 \ldots s$$

$$\lambda_j \geq 0, j = 1 \ldots N, S_i^-, S_r^+ \geq 0 \, \forall \, i \, and \, r, \ h_0 \text{ free.}$$

ε is a non-Archimedean infinitesimal.

As can be seen in model [M9.2] the non-discretionary input levels of DMU j_0 on the RHS of the respective constraints are multiplied by $\sum_{j=1}^{N} \lambda_j$.

This ensures that the most productive scale size yielded by the model in respect of DMU j_0 does not require larger levels on the non-discretionary inputs than those of DMU j_0.

To see this note that the MPSS levels yielded by [M9.2] in respect of the non-discretionary input levels of DMU j_0 are (see Chapter 6)

$$x_{ij_0}^{MPSS} = x_{ij_0} - \frac{S_i^{-*}}{\sum_{j=1}^{N} \lambda_j^*} \qquad i \in I_{nd} \qquad\qquad (9.1)$$

where * denotes the value of the respective variable at the optimal solution

to [M9.2]. The term $\dfrac{S_i^{-*}}{\sum\limits_{j=1}^{N} \lambda_j^*}$ cannot be negative.

Example 9.1: You manage six retail outlets selling clothes and wish to assess them on their efficiency in attracting custom (*market efficiency*). You decide to use DEA for this purpose and compile the data in Table 9.1.

Table 9.1.   Sales in ($m) and Floor Space (m²)

	Sales by nearest competitors	Sales by the outlet	Sales in the market served	Floor space
Outlet 1	20	3.5	35	110
Outlet 2	25	8	67	200
Outlet 3	14	11	108	220
Outlet 4	25	7	80	190
Outlet 5	28	4	35	120
Outlet 6	36	5	48	80

a)   Ignoring any issue of exogenously fixed inputs and outputs assess the market efficiency of outlet 4.

b)   Taking into account the fact that certain input and output variables are exogenously fixed set up a DEA model to assess the market efficiency of outlet 4.

c)   Solve the model in b) and comment on any differences in the solutions obtained in parts a) and b).

Solution:

a)   We need to identify the input and output variables and to decide on the most appropriate assumption we may maintain on returns to scale. Intuitively we would wish outlets to maximise sales within the competition and market size they face and allowing for their floor space. Thus, a sensible formulation would be to treat sales by each outlet as an *output* and the rest of the variables as *inputs*. However, competition is a non-isotonic input (see section 7 of Chapter 5), because larger levels of competition would lead to lower, not larger output, 'all else being equal'. There have been a variety of approaches for dealing with problems of this type (e.g. see Athanassopoulos and Thanassoulis (1995)) including inverting the level of competition and then using it as an input or using it as an output. We shall adopt the latter approach here and treat competition as an output. Thus our formulation is as follows:

Table 9.2.   Input-Output Variables for Assessing Market Efficiency

Inputs	Outputs
-Sales in the market served	-Sales by nearest competitors
-Floor space	-Sales by the outlet

The less restrictive assumption to make on returns to scale is that they are variable. It is difficult to justify constant returns to scale here since there is little reason to expect that under efficient operation scaling input variables by some constant will lead to the output variables also being scaled by that same constant.

Ignoring the issue of any variable being exogenously fixed we can in principle adopt either an input or an output orientation to assess pure technical efficiency. However, it is evidently more sensible to assess output efficiency here since the primary concern is to assess the extent to which sales by each outlet may be capable of being raised under improved market efficiency. Thus the model to assess the market efficiency of outlet 4 is as follows:

Max $Z + \varepsilon$ (SCOMP + SSALES + SMARKET + SFLR)   [M9.3]
Subject to:
$20L1 + 25L2 + 14L3 + 25L4 + 28L5 + 36L6 - 25Z - SCOMP = 0$:Comp. sales
$3.5L1 + 8L2 + 11L3 + 7L4 + 4L5 + 5L6 - 7Z - SSALES = 0$:Outlet sales
$35L1 + 67L2 + 108L3 + 80L4 + 35L5 + 48L6 + SMARKET = 80$:Market size
$110L1 + 200L2 + 220L3 + 190L4 + 120L5 + 80L6 + SFLR = 190$:Floor
$L1 + L2 + L3 + L4 + L5 + L6 = 1$
All variables non-negative, Z free. ε is a non-Archimedean infinitesimal.

The non-zero values at the optimal solution to this model are as follows:

Z	1.072368
SMARKET	6.934210
SFLR	51.513157
L3	0.417763
L6	0.582237

The pure technical efficiency of outlet 4 is $1 / Z = 1 / 1.072 = 0.9328$. The outlet attracts no more than 93.28% of the sales revenue it could in principle attract, against no more than 93.28% of the level of sales by competitors located nearest to the store that it could have in principle faced. Its efficient peers are outlets 3 and 6.

b) Clearly all input and output variables, bar sales revenue, are exogenously fixed. An outlet cannot influence demand for its products in the market in which it operates or the sales of its closest competitors nor does it in the short term control its floor space. (Strictly speaking the outlet may impact the sales of its closest competitors by its own market efficiency and indeed by the same token it may impact sales in the market in which it operates. We ignore, however, here such 'second order' impacts.) Implementing the output oriented version of the generic model in [M9.1] we solve the following model to assess the pure technical output efficiency of outlet 4.

Max $Z + \varepsilon SSALES$ [M9.4]
Subject to
$20L1 + 25L2 + 14L3 + 25L4 + 28L5 + 36L6 - SCOMP = 25$:Comp. sales
$3.5L1 + 8L2 + 11L3 + 7L4 + 4L5 + 5L6 - 7 Z - SSALES = 0$:Outlet sales
$35L1 + 67L2 + 108L3 + 80L4 + 35L5 + 48L6 + SMARKET = 80$:Market size
$110L1 + 200L2 + 220L3 + 190L4 + 120L5 + 80L6 + SFLR = 190$:Floor
$L1 + L2 + L3 + L4 + L5 + L6 = 1$
All variables non-negative, Z free, ε is a non-Archimedean infinitesimal.

c)   The non-zero values at the optimal solution to model [M9.4] are as follows:

Z	1.142757
L3	0.500000
L6	0.500000
SMARKET	2.000000
SFLR	40.000000

The pure technical output efficiency of outlet 4 is $1 / Z = 1 / 1.143 =$ 0.8748. The outlet attracts no more than 87.48% of the sales revenue it should attract given its floor space, sales by its closest competitors and sales in the market it serves. Its efficient peers remain outlets 3 and 6. (As we will see later there is an alternative optimal solution to [M9.4] in which outlets 2, 3 and 6 are efficient peers.)

The assessment of outlet 4 in b) is more realistic than that in a) because it measures the potential for expanding its sales controlling for the other three variables in the model which are beyond the outlet's control. It is not very meaningful to measure efficiency as we did in part a) in terms of how much higher competitor sales the outlet could have faced as the outlet can only influence such sales indirectly and probably only marginally.

Example 9.2: Repeat part (b) of Example 9.1 using the *Warwick DEA Software*.

Solution: The input file is
+compet +outletsales –marketsize -Floor
Outlet1 20 3.5 35 110
Outlet2 25 8 67 200
Outlet3 14 11 108 220
Outlet4 25 7 80 190
Outlet5 28 4 35 120
Outlet6 36 5 48 80
End.

We adopt a radial output orientated model. (See the User Guide in Chapter 10. Thus, the 'radial' priorities of inputs will be zero by default which is what we want since they are exogenous. All we need do is to also set their 'phase 2' priorities to zero. The radial priorities of output variables will all be by default equal to 100. To reflect the exogeneity of *competitor sales* its radial priory is set to zero. The same is done for its phase 2 priority. Figure 9.2 shows these settings. Upon running the software with the priorities in Figure 9.2 the log file obtained contains the following output in respect of Outlet 4:

Peers for Unit OUTLET4 efficiency 87.50% radial

OUTLET 4		OUTLET 2	OUTLET 3	OUTLET 6
ACTUAL	LAMBDA	0.800	0.100	0.100
80.0	-MARKET SIZE	67	108	48
190.0	-FLOOR	200	220	80
25.0	+COMPET	25	14	36
7.0	+OUTLET SALES	8	11	5

Targets for Unit OUTLET 4 efficiency 87.50% radial

VARIABLE	ACTUAL	TARGET	TO GAIN
-MARKET SIZE	80.0	69.2	13.5%
-FLOOR	190.0	190.0	0.0%
+COMPET	25.0	25.0	0.0%
+OUTLET SALES	7.0	8.0	14.3%

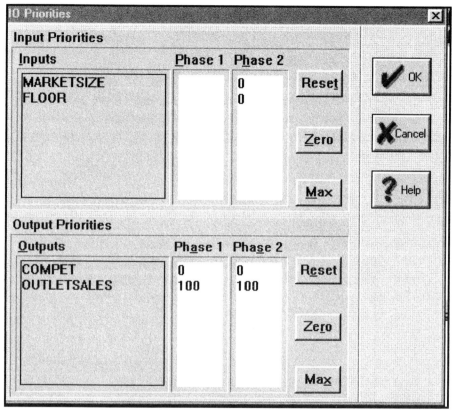

Figure 9.2.   Setting Radial and Phase 2 Priories to Reflect that Only OUTLETSALES
Are Not Exogenously Fixed

The efficiency measure of 87.5% matches what we obtained in Example
9.1. The *target* on market size merely reflects that Outlet 4 should be capable
of raising its sales by 14.3% even if the size of the market in which it
operates shrinks from $80m to about $69m.

9.3   IDENTIFYING PREFERRED PARETO-
EFFICIENT INPUT-OUTPUT LEVELS BY DEA

As we saw in Chapter 4 basic DEA models (e.g. [M4.1]) yield target
input-output levels which would render a DMU Pareto-efficient by giving
pre-emptive priority to the radial expansion of output levels, or alternatively,
to the radial contraction of input levels. (See section 4.6.1.) Such pre-
emptive priorities may not always accord with the path to efficiency a DMU
wishes to follow. The DMU management, or those in control of the DMU,
may wish to attain Pareto-efficiency giving varying priorities to
improvements in individual inputs and outputs.

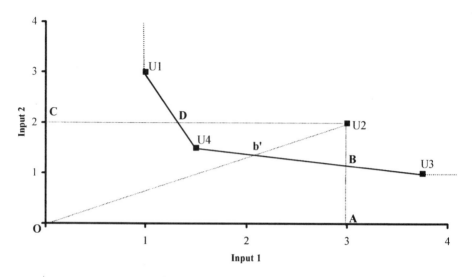

Figure 9.3.   Alternative Targets to Render DMU U2 Pareto-Efficient

Figure 9.3 illustrates the issue. It depicts a set of DMUs operating a CRS technology using two inputs to secure one output. Output has been standardised to one unit. The basic input-oriented DEA model [M4.1] yields targets at b'. However, were DMU U2 to give pre-emptive priority to the minimisation of input 1 the target input levels would be at D. Similarly, if pre-emptive priority is given to the minimisation of input 2 the target input levels would be at B. In fact when both inputs are under managerial control the targets for Pareto efficiency could in principle lie anywhere on the Pareto-efficient boundary U1U4U3. If we do impose the restriction that no input of DMU U2 should deteriorate when it attains Pareto-efficiency then its targets could lie anywhere on part DU4B of the Pareto efficient boundary.

Thanassoulis and Dyson (1992) developed two generic models for identifying Pareto-efficient targets by DEA in a manner whereby DM preferences over improvements to individual input and output variables are reflected. The first generic model restricts the targets so that no observed input or output level of the DMU deteriorates. The second generic model does permit deterioration of this kind within the target input-output levels.

9.3.1 Case Where no Input or Output May Deteriorate

Let us assume that we have N DMUs ($j = 1...N$) each one using m inputs to secure s outputs. Let us denote x_{ij} and y_{rj} the level of the ith input and rth output respectively observed at DMU j. Let us further assume that the DM partitions the inputs into subsets I_i and I_{nw}, containing inputs whose levels the DM wishes to see respectively improve and not worsen within the targets to be estimated. Finally assume that the DM similarly partitions the outputs into two subsets O_i and O_{nw}. Then a set of targets which would render DMU j_0 Pareto-efficient in line with DM preferences is $(x_i^t, i = 1 \ldots m, y_r^t, r = 1 \ldots s)$ where,

$$x_i^t = \sum_{j=1}^{N} \lambda_j^* x_{ij} = P_i^* x_{ij_0} \qquad\qquad i \in I_i$$

$$x_i^t = \sum_{j=1}^{N} \lambda_j^* x_{ij} = x_{ij_0} - S_i^{-*} \qquad\qquad I \in I_{nw} \qquad\qquad (9.2).$$

$$y_r^t = \sum_{j=1}^{N} \lambda_j^* y_{rj} = Z_r^* y_{rj_0} \qquad\qquad r \in O_i$$

$$y_r^t = \sum_{j=1}^{N} \lambda_j^* y_{rj} = y_{rj_0} + S_r^{+*} \qquad\qquad r \in O_{nw}$$

and λ_j^*, S_i^{-*}, S_r^{+*}, P_i^* and Z_r^* are respectively optimal values of λ_j, S_i^-, S_r^+, P_i and Z_r in model [M9.5].

$$\text{Max} \quad \sum_{r \in O_i} W_r^+ Z_r - \sum_{i \in I_i} W_i^- P_i + \varepsilon \left[\sum_{i \in I_{nw}} S_i^- + \sum_{r \in O_{nw}} S_r^+ \right] \qquad [M9.5]$$

Subject to:

$$\sum_{j=1}^{N} \lambda_j x_{ij} \leq P_i x_{ij_0} \qquad\qquad i \in I_i$$

$$\sum_{j=1}^{N} \lambda_j x_{ij} = x_{ij_0} - S_i^- \qquad\qquad i \in I_{nw}$$

$$\sum_{j=1}^{N} \lambda_j y_{rj} \geq Z_r y_{rj_0} \qquad\qquad r \in O_i$$

$$\sum_{j=1}^{N} \lambda_j y_{rj} = y_{rj_0} + S_r^+ \qquad\qquad r \in O_{nw}$$

$Z_r \geq 1 \; \forall \, r \in O_i$,
$P_i \leq 1 \; \forall i \in I_i$,
$S_r^+ \geq 0 \; \forall \, r \in O_{nw}$, $S_i^- \; \forall i \in I_{nw}$, $\lambda_j \geq 0 \; \forall \, j$.

W_i^- and W_r^+ are DM specified positive weights. Their magnitudes reflect the relative strength of preference the DM has for the improvement of the corresponding input or output. ε is a non-Archemedian infinitesimal.

The targets in (9.2) are Pareto-efficient. (The proof of this can be found in Thanassoulis and Dyson 1992.)

Note that model [M9.5] is akin to the *additive* DEA model (e.g. Cooper et al. (2000) Section 4.3) except that [M9.5] incorporates preferences over improvements to individual input and output variables and the preferences are expressed over proportional rather than absolute improvements. The objective function to the model is a weighted sum of proportional improvements to input-output variables and so the model does not yield an efficiency measure which can be readily interpreted. The purpose of the model is to estimate targets which would render a DMU Pareto-efficient in line with DM preferences rather than yield a measure of the distance of the DMU from the Pareto-efficient boundary.

In practice the DM is unlikely to articulate outright such values for the weights W_i^- and W_r^+ in [M9.5] that lead to targets s/he finds satisfactory. It is more likely the DM would need to try iteratively several sets of values of the weights W_i^- and W_r^+ before arriving at acceptable targets.

The model in [M9.5] can be readily modified for DMUs operating a VRS technology by simply adding the convexity constraint, requiring the λ's to sum to 1. The model can also be readily modified for the case where some of the input-output variables are exogenously fixed. Finally, one special case of model [M9.5] arises when the DM wishes to give pre-emptive priority to the improvement of one input or one output variable while wishing the observed values of the rest of the variables not to deteriorate within the targets to be determined. An illustration of the use of model [M9.5] in such a case can be found later in Example 9.3. The relevant modifications of model [M9.5] for all of the foregoing cases can be found in Thanassoulis and Dyson (1992).

9.3.2 Case Where Some Input or Output May Deteriorate

When we permit the observed input-output levels of a DMU to deteriorate we open, in principle, the entire Pareto-efficient boundary of the PPS as the locus of potential target input-output levels for the DMU. In such a case in order to arrive at a set of Pareto-efficient targets for a DMU that the DM may find satisfactory Thanassoulis and Dyson (1992) propose the use of *ideal targets* within a DEA framework. Ideal targets are user-specified input-output levels which the DM would in principle wish the DMU concerned to attain. In practice the ideal targets could be modifications to the observed input-output levels of the DMU, designed to reflect DM preferences over relative improvements in input-output levels. Ideal targets may incorporate a re-orientation of the observed mix of input-output levels of the DMU and deterioration to the observed levels of some variables for the benefit of improved levels in other variables. The DM is in no position to judge at the outset the feasibility of the ideal targets he/she proposes.

Once the ideal targets are specified a two-phase procedure is deployed to arrive first at a set of feasible and then at Pareto-efficient targets most compatible with the DM's preferences. Figure 9.4 illustrates the approach.

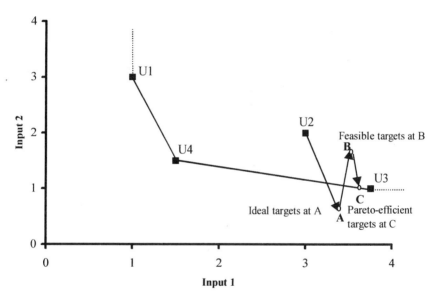

Figure 9.4.   Using DM Specified Ideal Targets to Arrive at Pareto-Efficient Targets

It depicts the same set of DMUs as Figure 9.3. Let us assume that the DM has specified ideal targets which lie at A. As point A (unbeknown to the DM) lies outside the PPS, the ideal targets are infeasible. It is therefore necessary to test the ideal targets for feasibility and, if necessary, 'move' them to a point within the PPS. The direction chosen for moving the targets should accord with the relative preferences the DM has for improving the levels of the individual inputs. The point we reach within the PPS may not necessarily be Pareto-efficient. For example the ideal targets may be moved for feasibility from A to B in Figure 9.4. It is therefore generally necessary to engage in a second phase to move from the point initially identified within the PPS to one which we know will be Pareto-efficient. Such a point could be C in Figure 9.4.

The procedure sketched in Figure 9.4 for arriving from ideal to Pareto-efficient targets within the PPS is implemented in the general case as follows. Let us assume that we have N DMUs ($j = 1...N$) each one using m inputs to secure s outputs. Let us denote x_{ij} and y_{rj} the level of the ith input and rth output respectively observed at DMU j. Let us further assume that the DM has specified ideal targets ($x_i^{id}, i = 1...m,\ y_r^{id}, r = 1...s$). Model [M9.6] yields feasible targets ($x_i^f,\ i = 1...m,\ y_r^f,\ r = 1...s$) compatible with the ideal targets and DM preferences where,

$$x_i^f = \sum_{j=1}^{N} \lambda_j^* x_{ij} \qquad\qquad i = 1...m$$

$$y_r^f = \sum_{j=1}^{N} \lambda_j^* y_{rj} \qquad\qquad r = 1...s \qquad\qquad (9.3)$$

and λ_j^*, is the optimal values of λ_j in [M9.6].

$$\text{Min} \qquad \sum_{i=1}^{m} w_i^{1-} c_i^{1-} + \sum_{i=1}^{m} w_i^{2-} c_i^{2-} + \sum_{r=1}^{s} w_r^{1+} k_r^{1+} + \sum_{r=1}^{s} w_r^{2+} k_r^{2+} \quad [M9.6]$$

$$\sum_{j=1}^{N} \lambda_j x_{ij} + c_i^{1-} - c_i^{2-} = x_{ij_0}^{id} \qquad\qquad i = 1...m$$

$$\sum_{j=1}^{N} \lambda_j y_{ij} + k_r^{1+} - k_r^{2+} = y_{rj_0}^{id} \qquad\qquad r = 1...s$$

$$\lambda_j, c_i^{1-}, c_i^{2-}\ k_r^{1+}, k_r^{2+} \geq 0\ \forall\ j,\ i\ \text{and r.}$$

Model [M9.6] is a *goal programming* model. The goals are the ideal target levels and the model estimates the feasible input-output levels ($x_i^f, i = 1...m,\ y_r^f, r = 1...s$) in (9.3) whose weighted distance from the ideal targets, reflected in the objective function to the model, is minimum. The weights w_i^{1-}, w_i^{2-}, w_r^{1+}, w_r^{2+} are user specified and they reflect the relative desirability of attaining the ideal levels of the respective inputs and outputs. Note that the weights used must take into account the scale of measurement of the input-output variables for otherwise the solution to the model will be dominated by those variables which are measured in large absolute values even when their weights in the objective function to [M9.6] may be relatively low.

The feasible input-output levels in (9.3) may not be Pareto-efficient. The model in [M9.7] (known as an *additive DEA model*) identifies any Pareto-efficient targets which may exist, dominating the feasible input-output levels in (9.3).

$$\text{Max} \qquad \sum_{i=1}^{m} d_i^- + \sum_{r=1}^{s} d_r^+ \qquad\qquad\qquad \text{[M9.7]}$$

$$\sum_{j=1}^{N} \alpha_j x_{ij} + d_i^{1-} = x_{ij_0}^f \qquad\qquad i = 1...m$$

$$\sum_{j=1}^{N} \alpha_j y_{ij} - d_r^+ = y_{rj_0}^f \qquad\qquad r = 1...s$$

$$\alpha_j, c_i^-, c_r^+ \geq 0 \; \forall \; j, i \text{ and } r.$$

Notation in model [M9.7] is as in [M9.6]. The Pareto-efficient targets yielded by [M9.7] are

$$x_i^{ef} = \sum_{j=1}^{N} \alpha_j^* x_{ij} \qquad\qquad i = 1...m$$

$$y_r^{ef} = \sum_{j=1}^{N} \alpha_j^* y_{rj} \qquad\qquad r = 1...s \qquad\qquad (9.4)$$

where α_j^*, is the optimal value of α_j in model [M9.7].

Example 9.3: Consider the mail order company of Example 4.1, Chapter 4.
a) Identify targets which would render Centre 1 Pareto-efficient giving equal priority to the reduction in labour used and the rise in number of packages delivered without detriment to the observed level of any one input or output variable.
b) Identify targets which would render Centre 1 Pareto-efficient giving pre-emptive priority to the reduction of labour used, without detriment to the observed level of any one input or output variable.
c) Repeat (a) above using the closest available model within *Warwick DEA Software*.

Solution:
a) The model in [M9.8], based on the generic model in [M9.5] can be solved to estimate the required targets. The value of 0.0001 is used to simulate the non-Archimedean infinitesimal ε in [M9.5].

Max $Z - P + 0.0001SC + 0.0001SD$ [M9.8]
Subject to:
Labour: $4.1P$ $\geq 4.1\lambda_1 + 3.8\lambda_2 + 4.4\lambda_3 + 3.2\lambda_4 + 3.4\lambda_5$
Capital: $2.3 - SC$ $= 2.3\lambda_1 + 2.4\lambda_2 + 2.0\lambda_3 + 1.8\lambda_4 + 3.4\lambda_5$
Distance: $4.3 + SD$ $= 4.3\lambda_1 + 3.9\lambda_2 + 4.1\lambda_3 + 5.2\lambda_4 + 4.2\lambda_5$
Packages: $90Z$ $\leq 90\lambda_1 + 102\lambda_2 + 96\lambda_3 + 110\lambda_4 + 120\lambda_5$
             $Z \geq 1$
             $P \leq 1$
             $\lambda_1, \lambda_2, \lambda_3, \lambda_4, \lambda_5, SC, SD \geq 0.$

This model yields $Z = 1.56$, $P = 0.997$, $SC = 0$, $SD = 2.34$. These values used within the expression in (9.2) lead to the following targets:

Centre 1	Labour Hours (000)	Capital employed ($m)	Delivery distance (000 km)	Packages delivered (000)
Observed levels	4.10	2.3	4.30	90.0
Targets Identified	4.09	2.3	6.64	140.4

b)   The input-output levels which would render Centre 1 Pareto-efficient giving pre-emptive priority to the reduction of labour without detriment to the rest of its input-output levels are obtained by solving the model [M9.9].

Min $Z - \varepsilon (SC+SP+SD)$ [M9.9]
Subject to:
Labour: $4.1Z$ $= 4.1\lambda_1 + 3.8\lambda_2 + 4.4\lambda_3 + 3.2\lambda_4 + 3.4\lambda_5$
Capital: $2.3 - SC$ $= 2.3\lambda_1 + 2.4\lambda_2 + 2.0\lambda_3 + 1.8\lambda_4 + 3.4\lambda_5$
Distance: $4.3 + SD$ $= 4.3\lambda_1 + 3.9\lambda_2 + 4.1\lambda_3 + 5.2\lambda_4 + 4.2\lambda_5$
Packages: $90 + SP$ $= 90\lambda_1 + 102\lambda_2 + 96\lambda_3 + 110\lambda_4 + 120\lambda_5$
             $\lambda_1, \lambda_2, \lambda_3, \lambda_4, \lambda_5, SC, SP, SD \geq 0.$

The solution to this model yields optimal values $Z = 0.645$, $SC = 0.81$, $SD = 0$, $SP = 0.96$. These values used within model [M9.9] lead to the following targets.

Centre 1	Labour Hours (000)	Capital employed ($m)	Delivery distance (000 km)	Packages delivered (000)
Observed levels	4.1	2.3	4.3	90
Targets Identified	2.644	1.49	4.3	90.96

c) The closest available model within *Warwick DEA Software* to model [M9.5] is that available as *Targets* model the rest of the settings being as under default. The input file for assessing the Centres by *Warwick DEA Software* is

-LABOUR -CAPITAL +DISTANCE +PACKAGES
C1 4.1 2.3 4.3 90
C2 3.8 2.4 3.9 102
C3 4.4 2 4.1 96
C4 3.2 1.8 5.2 110
C5 3.4 3.4 4.2 120
END

Once we have opened the input file we select the *Targets* model. Then we specify in the priorities dialog box (see the User Guide in Chapter 10) the phase 2 priorities as illustrated in Figure 9.5.

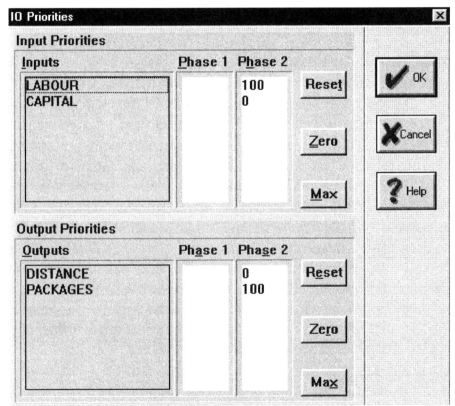

Figure 9.5. Giving Equal Priority to the Improvement of Labour and Packages Delivered

The targets yielded by Warwick DEA Software when the priorities in Figure 9.5 are applied to Centre 1 are as follows:

Centre 1	Labour Hrs(000)	Capital ($m)	Distance (000 km)	Packages (000)
Observed levels	4.10	2.3	4.30	90.0
Targets Identified	4.1	2.3	6.6	140.6

These targets are virtually the same as those in part (a) the differences being most likely the result of round-off errors. (The targets in this case happen to also coincide with those we would obtain if we solve in respect of Centre 1 the basic output oriented DEA model (model M4.3 Chapter 4).

9.4 ASSESSING DMUS IN THE PRESENCE OF CATEGORICAL VARIABLES

Categorical variables take discrete values. For example the categorical variable *location* of a DMU may take the value 1 for rural and 2 for urban location. If a categorical variable affects the productivity of DMUs then it is important that it should be reflected in their assessment. Two possibilities arise: We either *do* or *do not* maintain a prior assumption about the relative benefit a categorical variable value bestows on DMUs.

One example where we may maintain no prior assumption about a categorical variable and its impact on productivity would be whether police forces covering rural areas can inherently solve more readily crimes compared to forces covering urban areas. Where we maintain no prior assumption about the benefit a categorical variable bestows on DMUs we may wish to use DEA to investigate whether any such benefit exists, and if so, to identify any link between input-output mixes and productivity benefit the categorical variable bestows DMUs. The approach we can use in such cases is that for assessing *policy effectiveness*, covered in Chapter 7.

A case where we may maintain a prior assumption that a categorical variable bestows different productivity advantages depending on its value could be water treatment. A water treatment plant treating surface water would normally incur higher levels of expenditure than one treating ground water, all else being equal, because of the generally higher pollution of surface water. Where we do maintain a prior assumption of this kind we can use the approach below, developed by Banker and Morey (1986b), to assess the efficiencies of the DMUs. The central idea in the approach is that

DMUs which enjoy a higher productivity advantage by virtue of their value on a categorical variable may not be referent DMUs for less advantaged DMUs on that variable. However, a less advantaged DMU on a categorical variable may be referent for DMUs which are more advantaged on that variable.

This idea was operationalised by Banker and Morey (1986b) using an integer programming model. A simpler operationalisation was put forth by Charnes et al. (1994) relying on the solution of a sequence of linear programming models. We present the latter operationalisation here.

Let us assume that we have N DMUs ($j = 1$ N) using m inputs to secure s outputs and that their productivity is impacted by some categorical variable. Let us denote x_{ij} and y_{rj} the level of the ith input and rth output respectively observed at DMU j. Let us further assume that on the basis of the categorical variable the DMUs can be subdivided into mutually *exclusive* and *exhaustive* subsets S_1, S_2 ...S_L so that DMUs in subset i enjoy a productivity advantage over the DMUs in subsets 1, 2, ..., i-1. Then the technical input efficiency of DMU $j_0 \in S_p$, $p \le L$ is k_0^*, where k_0^* is the optimal value of k_0 in model [M9.10].

$$\text{Min} \qquad k_0 - \varepsilon \left[\sum_{i=1}^{m} S_i^- + \sum_{r=1}^{s} S_r^+ \right] \qquad \text{[M9.10]}$$

Subject to:
$$\sum_{\substack{j \in \bigcup_{k=1}^{k=p} S_k}} \lambda_j x_{ij} = k_0 x_{ij_0} - S_i^- \qquad i = 1...m$$

$$\sum_{\substack{j \in \bigcup_{k=1}^{k=p} S_k}} \lambda_j y_{rj} = S_r^+ + y_{rj_0} \qquad r = 1...s$$

$$\lambda_j \ge 0, \forall j, S_i^-, S_r^+ \ge 0 \forall i \text{ and } r, k_0 \text{ free.}$$

ε is a non-Archimedean infinitesimal.

This model permits only the DMUs which enjoy no better advantage on the categorical variable than DMU j_0 itself to feature as its efficient peers. Thus, if DMU j_0 is identified as Pareto-inefficient it cannot be on account of the value it has on the categorical variable.

The approach outlined here does not permit trade-offs between the level of productivity advantage the categorical variable bestows on a DMU and the rest of the input-output variables featuring within the assessment. The effect is that while any DMU found to be relatively inefficient cannot blame the value it has on the categorical variable, some DMUs may appear artificially more efficient than they really are. For example if the groups of the less advantaged DMUs on productivity are sparsely populated then their DMUs may appear more efficient than is justified, simply because of lack of suitable comparators.

The foregoing approach to handling categorical variables within DEA is applicable when the DMU has no control over the value of the categorical variable in its case. It is possible that the DMU has some control over the value of its categorical variable. For example it can opt to resource itself to offer a high or a medium level of service, representing two different values of the *service quality* categorical variable. Cooper et al. (2000, Chapter 7) offer a modification to the above approach applicable to cases of this type.

Example 9.4: Consider the mail order company in Example 4.1 in Chapter 4. Assume that Centres 1 and 2 serve urban environments, Centres 3 and 4 semi-urban and Centre 5 a rural environment. Assume further that due to traffic congestion for given labour and capital resources the urban environment is the least productive in terms of distance travelled and packages delivered. The semi-urban environment is more productive in this sense than the urban environment while the rural environment is the most productive of the three categories. Set up the DEA models to be solved to ascertain the technical input efficiency of each centre.

Solution: In the context of model [M9.10] we have $S_1 = \{1,2\}$, $S_2 = \{3,4\}$ and $S_3 = \{5\}$. The models needed are now set up for each subset of DMUs in turn.

a)   Urban environment (Centres 1 and 2)
The instance of the generic model [M9.10] corresponding to Centre 1 is as follows:

Min	$k - \varepsilon (SL + SC + SD + SP)$	[M9.11]
Such that:		
Labour:	$4.1k - SL$	$= 4.1\lambda_1 + 3.8\lambda_2$
Capital:	$2.3k - SC$	$= 2.3\lambda_1 + 2.4\lambda_2$
Distance:	$4.3 + SD$	$= 4.3\lambda_1 + 3.9\lambda_2$
Packages:	$90 + SP$	$= 90\lambda_1 + 102\lambda_2$

$\lambda_1, \lambda_2, SL, SC, SD, SP \geq 0$, k free,
ε is a non Archimedean infinitesimal.

The optimal value of k is the technical input efficiency of Centre 1.

Clearly in the case of Centre 2 the coefficients of k become 3.8 and 2.4 in the Labour and Capital constraints of [M9.11]. The LHS constants of 4.3 and 90 become respectively 3.9 and 102. The model remains otherwise as for Centre 1.

b)   *Semi-urban environment (Centres 3 and 4)*
The instance of model [M9.10] corresponding to Centre 3 is as follows:

Min	$k - \varepsilon (SL + SC + SD + SP)$		[M9.12]
Such that:			
Labour:	$4.4k - SL$	$= 4.1\lambda_1 + 3.8\lambda_2 + 4.4\lambda_3 + 3.2\lambda_4$	
Capital:	$2k - SC$	$= 2.3\lambda_1 + 2.4\lambda_2 + 2.0\lambda_3 + 1.8\lambda_4$	
Distance:	$4.1 + SD$	$= 4.3\lambda_1 + 3.9\lambda_2 + 4.1\lambda_3 + 5.2\lambda_4$	
Packages:	$96 + SP$	$= 90\lambda_1 + 102\lambda_2 + 96\lambda_3 + 110\lambda_4$	

$\lambda_1, \lambda_2, \lambda_3, \lambda_4, SL, SC, SD, SP \geq 0$, k free,
ε is a non-Archimedean infinitesimal.

The optimal value of k is the technical input efficiency of Centre 3.

Clearly in the case of Centre 4 the coefficients of k become respectively 3.2 and 1.8 in the Labour and Capital constraints of [M9.12]. The LHS constants of 4.1 and 96 become respectively 5.2 and 110. The model otherwise remains as for Centre 3.

d)   *Rural centres (Centre 5)*
The instance of model [M9.10] corresponding to Centre 5 is as follows:

Min	$k - \varepsilon(SL + SC + SD + SP)$		[M9.13]
Such that:			
Labour:	$3.4k - SL$	$= 4.1\lambda_1 + 3.8\lambda_2 + 4.4\lambda_3 + 3.2\lambda_4 + 3.4\lambda_5$	
Capital:	$3.4k - SC$	$= 2.3\lambda_1 + 2.4\lambda_2 + 2.0\lambda_3 + 1.8\lambda_4 + 3.4\lambda_5$	
Distance:	$4.2 + SD$	$= 4.3\lambda_1 + 3.9\lambda_2 + 4.1\lambda_3 + 5.2\lambda_4 + 4.2\lambda_5$	
Packages:	$120 + SP$	$= 90\lambda_1 + 102\lambda_2 + 96\lambda_3 + 110\lambda_4 + 120\lambda_5$	

$\lambda_1, \lambda_2, \lambda_3, \lambda_4, \lambda_5, SL, SC, SD, SP \geq 0$, k free,
ε is a non Archimedean infinitesimal.

The optimal value of k is the technical input efficiency of Centre 5.

9.5   QUESTIONS

(Where the context permits the reader may employ the Warwick DEA Software accompanying this book to answer the questions below. See Chapter 10 for a limited User Guide.)

1.   Reconsider the local tax offices of Question 4 in Chapter 4, section 8. The data for the most recent financial year is as follows.

Table 9.3.   Most Recent Data on Tax Offices

Office	Running Cost ($00,000)	Capital Charge ($0,000)	No of accounts (0,000)	Applns for adj'mnts (000)	No of Summonses (000)	NPV of taxes ($10m)
1	9.13	8.12	7.52	34.11	21.96	3.84
2	13.60	11.40	8.30	23.27	46.00	8.63
3	5.65	6.30	1.78	0.16	1.31	39.01
4	12.71	10.90	6.32	13.63	8.53	5.16
5	11.19	12.40	6.58	10.90	3.52	3.46

Assume that the offices operate a variable returns to scale technology. In respect of office 4:

i.   Identify a target level for operating expenditure, if its minimisation is to be given pre-emptive priority over improvements to the levels of the remaining input and output variables.

ii.   Identify a target level for accounts administered, if its maximisation is to be given pre-emptive priority over improvements to the levels of the remaining input and output variables.

iii.   Identify targets when ideally we would wish it to absorb no more than a 10% rise in output levels and lower its current levels of capital and operating expenditure as far as possible.

2. Consider the insurance salespersons in Table 9.4.

Table 9.4. Hours of Work and Financial Products Sold

Salesp'n	S1	S2	S3	S4	S5	S6	S7	S8	S9	S10
Hrs (00)	5	4.3	5	6	4.2	5.5	5.5	7.59	5.5	6.8
Life Policies	50	35	80	20	53	55	64	44.5	103	25
Pensions	30	40	35	50	25	33	38	54	44.5	64
Save Ac	45	50	40	30	35	74	57.5	60	51	38

Salesperson 8 operates in an environment where residents have a lower propensity to buy the products being sold than residents served by salespersons 9 and 10 but higher than those served by the rest of the salespersons.

i. Assess under constant returns to scale the technical output efficiency of salesperson 8 and identify targets that would render him/her Pareto-efficient.

ii. Identify targets that would render salesperson 8 Pareto-efficient if life policies and savings plans are to be given equal priority to improve and no priority is given to pension plans to improve or hours worked to reduce. Comment on any differences between these targets and those identified in part (i) above.

3. Six police forces are to be compared on their output efficiency using the following data:

Table 9.5. Input-Output Data for Comparing Police Forces

Force	Operating Expenditure (£m)	Crimes dealt with (000)	Crimes cleared (000)
1	1.2	52	10
2	1.4	50	9.5
3	2.2	86	18
4	1.8	79	14
5	1.7	50	13
6	2.5	105	16

Assuming variable returns to scale and treating the level of crimes dealt with as an exogenously fixed output estimate the technical output efficiency of Force 2 and identify targets that would render it Pareto-efficient. Comment on any differences between these targets and those that would be obtained when crimes dealt with are not treated as exogenously fixed.

4.   Reconsider the data on the six door-to-door representatives selling household improvements such as double glazing and building extensions, featured in Question 5 of Chapter 5, section 8. The data is reproduced below.

Table 9.6.   Data on Salespersons

Rep'tive	Hours of work (00)	Value of sales ($'000)	Households Signed Up
1	7.5	50	60
2	8	37	70
3	7	26	56
4	8.4	59	65
5	8.7	68	45
6	8.2	64	80

i.   Maintaining a constant returns to scale assumption identify a target for the value of sales salesperson 3 should attain if this is to be given pre-emptive priority over the number of households signed up and the salesperson is not to reduce hours of work. Comment on any difference between this target and that obtained when value of sales and number of households signed up have equal priority to improve and hours of work are to stay fixed.

ii.   Repeat i) if salesperson 3 operates in a more productive environment than salespersons 1 and 2 but no more so than the environments served by the remaining salespersons. Comment on any differences between the target for value of sales in (i) and the corresponding target for value of sales identified in (ii).

5.   Let us consider DMUs ($j = 1...N$) using m inputs to secure s outputs. Let us denote x_{ij}, and y_{rj} the level of the ith input and rth output respectively at DMU j. Let us further assume that the outputs can be partitioned into two subsets O_d, and O_{nd} consisting respectively of *non-discretionary* (i.e. exogenously fixed) and *discretionary* (i.e. under managerial control) outputs. All inputs are discretionary.

i.   Write a DEA model which can be used to assess the pure technical output efficiency of DMU j_0, assuming that the DMUs operate a VRS technology.

ii.   Repeat i) assuming the DMUs operate a CRS technology.

Chapter 10

A LIMITED USER GUIDE FOR *WARWICK DEA SOFTWARE*

10.1 INTRODUCTION

This book is accompanied by a limited version of *Warwick DEA Software*. The author is grateful to the University of Warwick, Coventry CV4 7AL in England for permission to incorporate the software with the book. The version is limited and capable of assessing only 10 DMUs. Please note:

-   The software accompanying the book is subject to the restrictions within the license document included with the software files.

-   Additional guidance on using the software can be obtained using the **Help** menu of the software.

Your have been supplied with these files:

Deawin.exe: The DEA program, for Windows 3.1 or higher.
Deawin.help: The help DEA file.
Licence.doc: File containing the terms and conditions under which the software is provided.
Banking.dat: An example data set relating to 10 bank branches.
Banking.log: An example log file of using *Warwick DEA Software* on Banking.dat.
Readme.txt: File with instructions for installing the software.

Please begin by reading the *Readme.txt* file for directions on making a copy of the software for safekeeping and installing it on your machine.

The instructions below take you from the point where you have made a copy of the software on floppy disk or have installed it or your machine.

You will need:

- An IBM 386 or higher or 100% IBM-Compatible computer with a minimum of 4 MB of RAM.

- Windows 3.1 (or higher).

- A hard disk drive with at least 2 MB of disk space.

10.2 PREPARING YOUR DATA INPUT BEFORE YOU INVOKE THE PROGRAM

Once you have identified the input-output variables you wish to use and the DMUs you wish to assess create a file with the input-output data of the units to be assessed. We shall refer to this as the *input* file. You can create the input file using an editor such as *Microsoft Word* or a spreadsheet such as *Microsoft Excel*. Bear in mind the following points:

⇒ The input file must have one of the two structures outlined under Data Formats below.

⇒ The input file must be an ASCII text file. So if you use an editor or a spreadsheet you should save the file unformatted, i.e. **text only**.

Two formats are available for the input file. The default format has input and output variables as columns and the DMUs as rows, e.g.:

	-COST	-MEN	+CARS	+LORRIES	+VANS
UNIT1	7.0	4.0	26.0	15.0	30.0
UNIT2	8.0	9.0	25.0	17.0	28.0
UNIT3	6.0	5.5	32.0	19.0	19.0
UNIT4	6.0	4.0	39.0	18.0	19.5
END					

An alternative format is the transpose of the above, with units as the columns and input-output variables as the rows. e.g.:

	UNIT1	UNIT2	UNIT3	UNIT4
+ CARS	26.0	25.0	32.0	39.0
-COST	7.0	8.0	6.0	6.0
+ LORRIES	15.0	17.0	19.0	18.0
-MEN	4.0	9.0	5.5	4.0
+ VANS	30.0	28.0	19.0	19.5
END				

The file is terminated by the word END. The names used must start with a letter, may contain up to 12 characters but must not include spaces or slashes. The + and - signs are optional. If a variable is to be an input it must be preceded by the - sign and if it is to be an output it must be preceded by the + sign. The signs can be overridden in that variables can be redesignated as input or output at run time. Each new row of a data file should start on a new line, as should the word END. Apart from this restriction, the spacing and general layout of data files is not of importance, and in most cases each row will require a number of lines within the file.

10.3 INITIATING A RUN OF THE PROGRAMME

Once you have saved and closed the input file you can initiate a run of the programme as follows:

If you have installed the programme double-click the DEA icon. If you have only made a copy on floppy disk open the folder containing the programme and double click the file **Deawin.exe**.

To terminate a session click File and then Exit.

To load the input file proceed as follows:

1. From the <u>File menu</u> choose **Open...** .
2. Select the data file you want in the Filename dialog box.
3. Click **✔ OK**.

The software now attempts to create **in the folder from which it read the input file** a log file by the same name as the input file but using the extension .log. If the log file already exists in that folder it will ask you whether you wish another log file name. If you choose **✔ Yes** you will be asked to enter the new filename. If you choose **🚫 No** then you may append to the current log file.

10.4 OPTIONS MENU

Improvement Model

You can choose **<u>Radial</u>**, **<u>Targets</u>** or **<u>Mixed</u>**.

Radial models

If you choose radial you may then choose: <u>Inputs</u>, <u>Outputs</u> or <u>Both</u>.

Inputs (Minimisation) (Default setting)

This is the default setting and corresponds to solving Model [M4.1] in Chapter 4. A two-phase process is followed minimising first the radial contraction of input levels and then maximising the sum of slacks. (See the description of model [M4.1] in Chapter 4.)

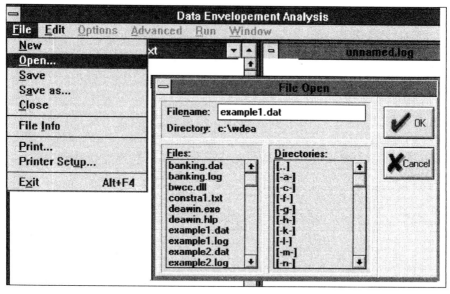

Outputs (Maximisation)

This setting corresponds to solving model [M4.3], in Chapter 4. Again a two-phase solution process is followed maximising first the radial expansion of output levels and then maximising the sum of slacks.

Both

Specialised model not covered in this guide but see the help file.

Targets model

The *Targets* model is close in spirit to model [M9.5] in Chapter 9 but not identical to it. The software solves model [M10.1]. Notation in [M10.1] is as in [M9.5], except that we no longer have a distinction between inputs (or outputs) to be improved and those not permitted to deteriorate. W_i^- and W_r^+ are user-specified positive weights. Their magnitudes reflect the relative strength of preference the user has for the improvement of the corresponding input or output. It is these weights that the user specifies as *phase 2 priorities* (see priorities dialog box below) but the software scales the weights the user provides to allow for any differences in magnitude between the different input and output variables. Notation in [M10.1] is as in [M9.5].

$$\text{Max} \quad \sum_i W_i^- S_i^- + \sum_r W_r^+ S_r^+]$$ [M10.1]

Subject to:

$$\sum_{j=1}^{N} \lambda_j x_{ij} = x_{ij_0} - S_i^- \qquad i = 1\ldots m$$

$$\sum_{j=1}^{N} \lambda_j y_{rj} = y_{rj_0} + S_r^+ \qquad r = 1\ldots s$$

$$\lambda_j \geq 0, j = 1 \ldots N, S_i^-, S_r^+ \geq 0 \, \forall \, i \, and \, r.$$

Mixed model

This model is aimed for identifying Pareto-efficient targets by a mixture of radial and non-radial improvements to inputs and outputs. The details are beyond the scope of this limited User Guide but are available with the full User Guide. See also the help menu.

Returns to Scale

Constant:
This is the default setting.

Variable:
Under this option when the *Radial Improvement Model* is used with *input minimisation* the software solves model [M6.8] while with *output maximisation* the software solves model [M6.9] in Chapter 6. If DMU j_0 turns out to be Pareto-efficient the software also solves models to identify the range of ω and w values in [M6.8] and [M6.9] respectively compatible with the Pareto-efficiency of DMU j_0. The range is reported within the **weights table** (see below) as *omega range* (Ω) in the case of both models. The range of Ω values is interpreted as follows:

- If the range is positive increasing returns to scale hold at the part of the efficient boundary where DMU j_0 is located;

- If the range is negative decreasing returns to scale hold at the part of the efficient boundary where DMU j_0 is located;

- If the range includes 0 constant returns to scale hold at the part of the efficient boundary where DMU j_0 is located.

To make this interpretation possible for both input minimisation and output maximisation models, the software negates appropriately before printing the Ω range it identifies.

Own Unit as Comparator

The user may select *'include'* or *'exclude'*. The default is 'include'. The options work as follows:

Include:

All DMUs, including the one being assessed, may feature as efficient comparators.

Exclude:

All DMUs, *except the one being assessed*, may feature as efficient comparators. Thus this setting solves models for estimating so called 'super-efficiency' (Andersen and Petersen (1993)).

Input-output priorities dialog box

The full details of how the priories here relate to the models solved by *Warwick DEA Software* are beyond the scope of this limited user guide. In brief, under the input minimising radial model, whether under constant or variable returns to scale, the phase 1 output priorities cannot be altered from the default of zero. Similarly, under the output maximising radial model, whether under constant or variable returns to scale, the phase 1 input priorities cannot be altered from the default of zero. In both cases the positive priorities are 100 and reflect equal preference to the contraction of input levels or expansion of output levels in the input minimising and output maximising setting respectively. We can change the phase 1 default radial preferences of 100 to zero to handle exogenously fixed (non-discretionary) variables as explained below.

The phase 2 priorities can be altered at will and can be thought of as the coefficients of the slacks in models [M4.1] or [M4.3] Chapter 4 when the setting relates to *radial improvement models*. The use of phase 2 priorities with *both, mixed* or *targets* models was briefly illustrated in Chapter 9 but the full details are beyond the scope of this limited user guide.

To set priorities select **Options** and click **Priorities** then **√ Radial** . A screen such as that below appears. Set the first and/or second phase priorities as desired either to reflect the relative desirability of improving input and output levels and/or to reflect non-discretionary variables. 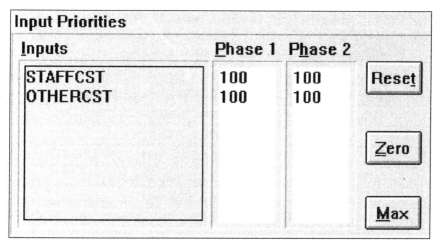 restores the priority entries to their initial values. sets all priority entries to zero. sets all priority entries to 100. Click when the desired priorities are set. Click to abort. Click to bring up the help screen.

Input Priorities

Inputs	P<u>h</u>ase 1	P<u>h</u>ase 2	
STAFFCST	100	100	Rese<u>t</u>
OTHERCST	100	100	
			<u>Z</u>ero
			<u>M</u>ax

Output Priorities

<u>O</u>utputs	Ph<u>a</u>se 1	Pha<u>s</u>e 2	
DEPAC	100	100	R<u>e</u>set
LOANS	100	100	
NEWAC	100	100	
			<u>Z</u>ero
			Ma<u>x</u>

Non-discretionary variables

We saw in Chapter 9 the Banker and Morey (1986a) models for assessing efficiency when certain input or output variables are exogenously fixed. The models relating to the case where efficient production is characterised by

variable returns to scale can be solved using *Warwick DEA Software* as follows:

Input non-discretionary variables

Invoke the *Radial Improvement Model, Input Minimisation.* Using the priorities dialog box set the **radial** or phase 1 priorities to zero for those input variables that are non-discretionary.. The model solved will be model [M9.1] in Chapter 9.

Output non-discretionary variables

Invoke the *Radial Improvement Model, Output Maximisation.* Using the priorities dialog box set the **radial** or phase 1 priorities to zero for those output variables that are non-discretionary. The model solved will be model [M9.1] in Chapter 9, re-oriented to output maximisation.

Note that the efficiency measure yielded in this case does <u>not</u> take into account feasible improvements to non-discretionary variables.

Efficiency

For the **radial** input minimisation or output maximisation models with uniform radial priorities, constant returns to scale the software reports respectively the technical input and the technical output efficiency of the DMU being assessed. This also holds true for variable returns to scale but the efficiencies are pure technical. The efficiencies reported when radial priorities are <u>not</u> uniform or the model used is not radial the efficiency rating reported does not have a radial contraction or expansion interpretation. (For more details on the efficiency reported see the help menu.)

10.5 THE RUN MENU

Unit selection

All DMUs are automatically selected, that is to say they are active, when the data is first loaded. It may be necessary, however, to change the active set interactively. To select units click **Run** and then select **Select Units Ctrl+U**. A screen such as that below appears.

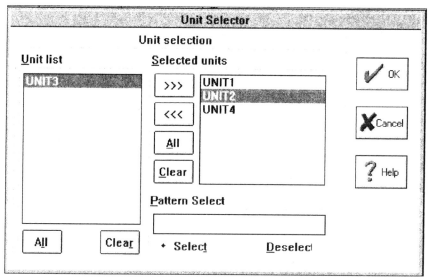

The Unit list shows all units that are not currently active while the Selected units shows all active units. Highlight a unit or set of units and then use <<< or >>> to transfer the selected units to the Unit list or to the Selected units respectively. Click ✓ OK when all active units are set. Click ✗ Cancel to abort. Click ? Help to bring up this screen again.

Input and output selection

In most cases, inputs and outputs are specified using prefix of + or - to variable names, when the data set is first loaded from disk. It may be necessary, however, to change them interactively.

To select input-output variables choose **Run** then select **Select IOs   Ctrl+I**. You will find an I/O selector window appears. E.g.:

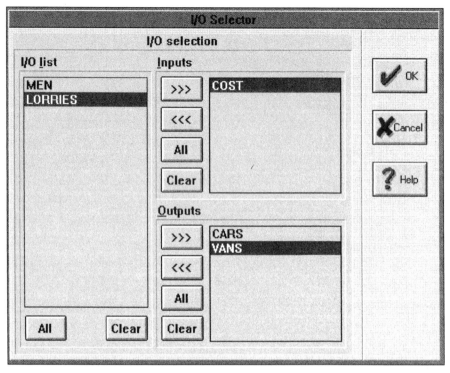

The I/O list shows all variables that were not specifically set to be inputs or outputs, Inputs shows all active input variables and Outputs contains all active output variables. Highlight a variable or set of variables and then use [<<<] or [>>>] to transfer the selected variables to or from the I/O list. Click [✓ OK] when the desired inputs and outputs have been selected. Click [✗ Cancel] to abort. Click [? Help] to bring up this screen again.

Execute dialog box

Before running the optimiser at least one variable must be designated as input and another as output. Furthermore, at least one unit must be selected. To run the optimiser select **Run** then **Execute   Ctrl+R**. The execute dialog described next appears.

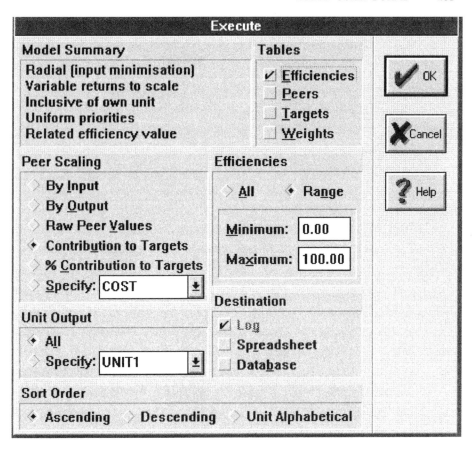

Model Summary reminds you what model you have set up. The <u>Tables</u> section controls the way the results are reported. There are four main types of Table:

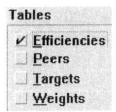

One or more of the selections can be active at any one time. The ordering of the units in these tables depends on the <u>Sort Order</u> selected.

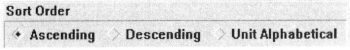

Partial reports can also be generated. The <u>Efficiencies</u> section controls the range of efficiencies to be reported (this is by default 0% to 100%), while the <u>Unit Output</u> section allows reporting individual units.

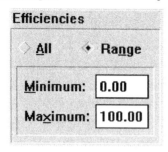

The <u>Peer Scaling</u> section controls the relative scaling of peer data to be reported. An illustration of how peer scaling can be of practical value in comparing an inefficient DMU with its efficient peers was given in Chapter 5 within the illustrative use of *Warwick DEA Software* on five bank branches. It is suggested, however, that Raw Peer values be used with the accompanying limited version of the software. Selecting *"Raw Peer Values"* would mean that the unscaled input-output levels of the efficient peer DMUs are printed. (The details of all peer scaling options are beyond the scope of this limited User Guide.)

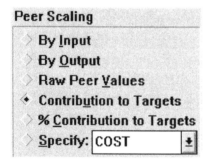

You can print the results for all DMUs or for a specific DMU. E.g. the selection below prints the results for all DMUs.

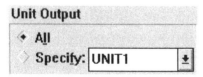

Destination files

The log file is created automatically and contains all output from a given session with the software. The Destination section permits the user to create additionally spreadsheet or database compatible files.

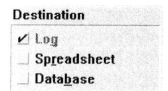

When the desired options are set, click to run the optimiser. The files printed with the Spreadsheet setting have the following formats.

DEA Efficiencies (Spreadsheet) File Format

The extension of this file is: *.tef and the format is as follows:

```
UNIT,RADIAL
UNIT2,87.85
UNIT1,100.00
UNIT3,100.00
UNIT4,100.00
****
```

DEA Lambda's (Spreadsheet) File Format

The extension of this file is *.tlf and the format is as follows:

```
UNIT,PEER,LAMBDA
UNIT2,UNIT1,0.733
UNIT2,UNIT3,0.316
UNIT1,UNIT1,1.000
UNIT3,UNIT3,1.000
UNIT4,UNIT4,1.000
****
```

UNIT1 and UNIT3 are peers to UNIT2 with the lambda values shown.

DEA Peers (Spreadsheet) File Format

The extension of this file is *. tpf and the format is as follows:
UNIT,PEER,COST,MEN,CARS,LORRIES,VANS
UNIT2,UNIT1,5.1,2.9,19.1,11.0,22.0
UNIT2,UNIT3,1.9,1.7,10.1,6.0,6.0
UNIT1,UNIT1,7.0,4.0,26.0,15.0,30.0
UNIT3,UNIT3,6.0,5.5,32.0,19.0,19.0
UNIT4,UNIT4,6.0,4.0,39.0,18.0,19.5
****

The data is of the peer and is scaled as selected under 'peer scaling' in the 'Execute' window.

DEA Targets (Spreadsheet) File Format

The extension of this file is *. ttf and the format is as follows:
UNIT,COST_A,COST_T,MEN_A,MEN_T,CARS_A,CARS_T,LORRIES_A,LOR
RIES_T,VANS _A,VANS_T
UNIT2,8.0,7.0,9.0,4.7,25.0,29.2,17.0,17.0,28.0,28.0
UNIT1,7.0,7.0,4.0,4.0,26.0,26.0,15.0,15.0,30.0,30.0
UNIT3,6.0,6.0,5.5,5.5,32.0,32.0,19.0,19.0,19.0,19.0
UNIT4,6.0,6.0,4.0,4.0,39.0,39.0,18.0,18.0,19.5,19.5
****

Data is in the order in which variable names appear above. The extension _A indicates actual (i.e. observed) data. The extension _T indicates target (i.e. efficient) value.

DEA Virtual (Spreadsheet) File Format

The extension of this file is *. tvf and the format is as follows:
UNIT,COST_V,COST_W,MEN_V,MEN_W,CARS_V,CARS_W,LORRIES_V,LO
RRIES_W,VA, NS_V,VANS_W
UNIT2,100.00,0.12500,0.00,0.00000,0.00,0.00000,35.04,0.02061,52.81,0.01886
UNIT1,22.69,0.03241,77.31,0.19329,22.69,0.00873,22.69,0.01512,54.63,0.01821
UNIT3,91.55,0.15259,8.45,0.01536,8.45,0.00264,83.11,0.04374,8.45,0.00445
UNIT4,66.67,0.11111,33.33,0.08333,33.33,0.00855,33.33,0.01852,33.33,0.01709
****

The extension _W indicates DEA weight. The extension _V indicates virtual level. Note that the software uses the normalisation

$$\sum_i \text{INPUT}_i \times \text{DEA INPUT WEIGHT}_i = 1$$

in the input minimisation orientation.   The normalisation in the output orientation is analogous.

10.6 THE ADVANCED MENU

This dialog box is for entering weights restrictions. You can enter weights restrictions either interactively or by reading them from a prepared file. We present the former case here for the limited version of the software accompanying this book. Select **Advanced** from program menu and then **Weight Restrictions**. A window such as the following will appear.

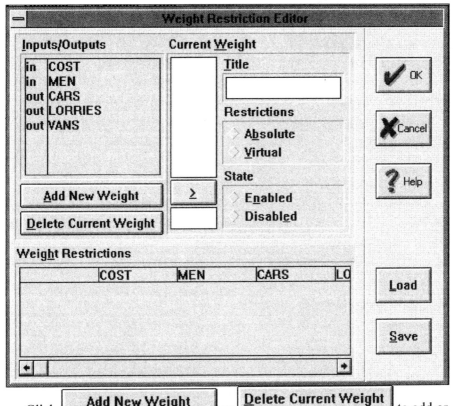

Click 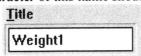 or **Delete Current Weight** to add or remove a constraint respectively. Type a name for the constraint in the Title box if adding a new restriction or changing the name of one. The first character of this name should be a letter.

Title

Weight1

Set the Absolute or Virtual restriction depending on whether you wish to restrict the DEA weights or the virtual (product of the DEA weight with the corresponding input or output level).

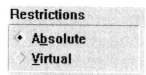

Then <u>Enable</u> the constraint. At anytime you can disable a constraint or

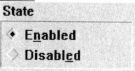

enable it.

Enter the weight and select <u>≥</u> or <u>≤</u> in the space to the left of Disabled. The weight relation can toggle between ≥ and ≤. Inputs and outputs are denoted with *in* and *out* on the left column respectively. Variables designated as neither are ghosted out. Use the *tab* key (or *shift tab*) to move between individual weights.

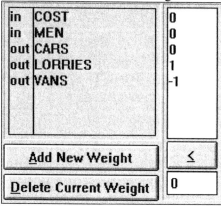

If you want to save the current constraint(s) in a file click <u>Save</u>.
The following part is a list of enabled and disabled constraint(s).

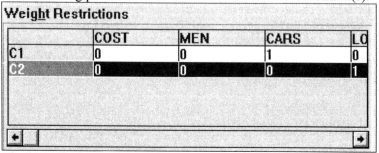

The box displays an overall view of the weight restriction set. It is also the only way to change individual constraints. The *Current Weight* section of the dialog box always displays the constraint selected.

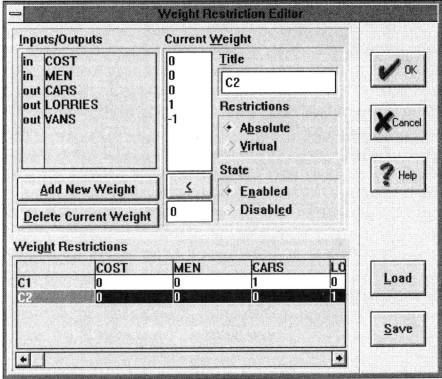

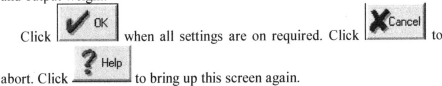

Editing an individual constraint involves changing the <u>constraint title</u>, it's <u>type</u>, whether it is <u>enabled</u> or not and of course the coefficient for each input and output weight.

Click [OK] when all settings are on required. Click [Cancel] to abort. Click [Help] to bring up this screen again.

Note that:
- DEA weights restrictions can render a DEA model infeasible;
- Weights restrictions under variable returns to scale are not supported by the software.

10.7 FILE HANDLING

Changes to the log files are saved automatically. You can save the input and log files using different names to the input file name. For this select **Save as...** from <u>File menu</u> then select directory and type your new file name in the Filename dialog box. To print your input or log file to the default printer select **Print...** from <u>File menu</u>.

You can import the 'spreadsheet' files created by *Warwick DEA Software* into a spreadsheet such as *Microsoft Excel* for further manipulation. Similarly you can open the log files created by *Warwick DEA Software* for further editing.

10.8 SOME COMMON ERROR MESSAGES

Could not open data window.
Too many applications are open and there is not enough real and/or virtual memory available. Close some applications and try again.

Failed to create main window.
Too many applications are open and there is not enough real and/or virtual memory available. Close some applications and try again.

Could not open (log) file!
The file is very probably locked by another application or mis-specified.

Could not create (log) file!
There is insufficient space on the destination drive for the new file, or the drive is write-protected.

Invalid file format. Numeric value was expected. Continue?
The file does not appear to be a DEA data file. It contains non-numeric data cells in other than the first row or column.

Premature End Of File. The file does not appear to be a DEA data file. Encountered the end-of-file mark before reading was completed. Such a file may have been created by an interrupted save operation. Note that this is a warning. The loader will in most cases recover by deleting the last line.

Could not open (create) log window.
Too many applications are open and there is not enough real and/or virtual memory available. Close some applications and try again.

Not enough memory for linear engine! Close some applications and try again.
Too many applications are open and there is not enough real and/or virtual memory available. Close some applications and try again.

Invalid matrix size! You MUST select some inputs/outputs before running.

Running, setting IO priorities and/or weight restrictions requires at least one input and one output to be selected.

<u>1st and/or 2nd phase priorities where NOT set. No IOs were selected.</u>
Running, setting IO priorities and/or weight restrictions requires at least one input and one output to be selected.

Author Index

Topic Index

References

Allen R. and E. Thanassoulis (1996) Increasing Envelopment in Data Envelopment Analysis. *Working Paper No. 210,* Warwick Business School, Warwick University, Coventry CV4 7AL, UK.

Allen R. (1997) *Incorporating Value Judgements in Data Envelopment Analysis.* PhD thesis, Warwick Business School, Warwick University, Coventry, CV4 7AL, England

Andersen, P., and Petersen, N. (1993), A Procedure for Ranking Efficient Units in Data Envelopment Analysis, *Management Science* Vol. 39 No. 10, pp. 1261-1264

Anderson D. R., D. J. Sweeney and T. A. Williams, (1998) *Quantitative Methods for Business,* (7th ed.) South-Western College Publishing.

Appa G. and Yue M. (1999) On Setting Scale Efficient Targets in DEA, *Journal of the Operational Research Society,* Vol. 50 No. 1, pp. 60-69.

Athanassopoulos A. D. and E. Thanassoulis (1995) Separating market efficiency from profitability and its implications for planning. *Journal of the Operational Research Society,* Vol. 46, No. 1. pp. 20 -34.

Banker, R.D. (1984), Estimating most productive scale size using data envelopment analysis, *European Journal of Operational Research,* Vol.17, pp. 35-44.

Banker, R.D., Charnes, A., and Cooper, W.W. (1984), Some models for estimating technical and scale inefficiencies in Data Envelopment Analysis , *Management Science* 30, 1078-1092.

Banker R. D. and R C Morey (1986a) Efficiency Analysis for Exogenously Fixed Inputs and Outputs *Operations Research* Vol. 34 pp. 513 - 521.

Banker R. D. and R C Morey (1986b) The Use of Categorical Variables in Data Envelopment Analysis, *Management Science* 32 (12) pp. 1613-1627.

Banker R. D and Thrall R. M. (1992) Estimation of Returns to Scale Using Data Envelopment Analysis, *European Journal of Operational Research* 62 pp. 74-84.

Berger, A.N. and Humphrey, D.B. (1997), Efficiency of financial institutions: International survey and directions for future research, *European Journal of Operational Research*, Vol. 98, pp. 175-212.

Carrington R., Puthucheary N. Rose D. (1997) Performance Measurement in Government Service Provision: The case of police services in New South Wales, *Journal of Productivity Analysis* Vol. 8, No 4 415-430.

Charnes A., W. W. Cooper (1962) Programming With Linear Fractional Functionals, *Naval Logistics Research Quarterly*, 9 (3,4) pp. 181-185.

Charnes, A., Cooper, W. W., Lewin, Y. A., and Seiford, M.L. (Eds) (1994), *Data Envelopment Analysis: Theory, Methodology and Application*, (Kluwer Academic Publishers).

Charnes A., W. W. Cooper and E. Rhodes (1978) Measuring the efficiency of decision making units, *European Journal of Operational Research* 2, 429-444

Charnes, A., W., Cooper and E., Rhodes, (1981), Evaluating Program and Managerial Efficiency: An Application of Data Envelopment Analysis to Program Follow Through , *Management Science*, Vol. 27, No. 6, pp. 668-697.

Coelli T., D.S. Prasada Rao, G. E. Battese (1998) 'An introduction to Efficiency and Productivity Analysis' Kluwer Academic Publishers Boston/ Dordrecht/London.

Cooper W. W., K. S. Park and J. T. Pastor Ciurana (2000b) Marginal Rates and Elasticities of Substitution with Additive Models in DEA, *Journal of Productivity Analysis* Vol. 13, No. 2 pp. 105-123

Cooper W.W., L. M. Seiford and K. Tone, (2000) *Data Envelopment Analysis: A comprehensive text with models, applications, references and DEA-solver software,* Kluwer Academic Publishers

Dyson R. G. (2001) Performance Measurement and Data Envelopment Analysis-rankings are ranks! *OR Insight*, Vol. 13, No. 4 pp3-8.

Dyson R. G., R. Hurrion and E. Thanassoulis (1993), Operational Research, Study Notes, Distance Learning MBA, Warwick Business School, Warwick University, Coventry CV4 7AL, UK.

Dyson R. G. and E. Thanassoulis (1988), Reducing Weight Flexibility in Data Envelopment Analysis. *Journal of the Operational Research Society*, Vol. 39, No. 6. pp. 563-576.

Färe, R. and Grosskopf, S. (1994) Understanding the Malmquist productivity index, Discussion Paper No 94-5, Southern Illinois University, Illinois

Färe, R., Grosskopf, S., Lindgren, B. and Roos, P. (1989). *Productivity developments in Swedish Hospitals: A Malmquist Output Index Approach.* Discussion Paper No 89-3, Southern Illinois University, Illinois.

Fare R.; Grosskopf, S.; and Lovell, Knox C.A. (1994), Production Frontiers, Cambridge University Press, Cambridge (England), New York (USA), Melbourne (Australia).

Farrell, M.J. (1957), The Measurement of Productive Efficiency , *Journal of the Royal Statistical Society*, Series A (general) 120, Part 3, 253-281

Golany, B. and Storbeck, J. (1999), A Data Envelopment Analysis of the Operational Efficiency of Bank Branches. *Interfaces* Vol. 29, No 3, pp. 14-26.

Kumbhakar, S. C. and Lovell, C.A. K. (2000) *Stochastic Frontier Analysis*, (Cambridge University Press)

Lang P., O.R. Yolalan and O. Kettani (1995). Controlled Envelopment by Face Extension in DEA. *Journal of the Operational Research Society,* 46, 4, 473-491

Maniadakis N. and E. Thanassoulis (forthcoming) A Cost Malmquist Productivity Index, *European Journal of Operational Research.*

Maniadakis N., and E. Thanassoulis (2000) Changes in Productivity of A Sample of Scottish Hospitals: A Cost Index Approach, *Applied Economics*, Vol. 32 pp. 1575-1589.

OFWAT (1995), *South West Water Services Ltd* , HMSO.

Olesen, O.B. and Petersen, N.C. (1996), Indicators of Ill-conditioned Data Sets and Model Misspecification in Data Envelopment Analysis: An Extended Facet Approach, *Management Science,* Vol. 42, No. 2, 205-219.

Podinovski V. (1999) Side effects of Absolute Weight Bounds in DEA Models, *European Journal of Operational Research* 115 pp. 583-595.

Roll Y. and B. Golany (1993). Alternate Methods of Treating Factor Weights in DEA. *OMEGA* Vol. 21, No. 1, pp. 99-109.

Roll Y., W.D. Cook and B. Golany (1991), Controlling Factor Weights in DEA. *IIE Trans.* 23, 1, 2-9.

Sarrico C. S. *Performance Measurement in UK Universities: Bringing in the Stakeholders' Perspectives Using Data Envelopment Analysis*, PhD Thesis, Warwick Business School, University of Warwick Coventry CV4 7AL, UK.

Sherman, H. D. and Ladino, G. (1995), Managing Bank Productivity Using Data Envelopment Analysis (DEA), *Interfaces* Vol. 25, No.2, pp. 60-73.

Smith, P. (1995) On the Unintended Consequences of Publishing Performance Data in the Public Sector , *International Journal of Public Administration*, 18 (2/3), pp. 277-310

Thanassoulis E., (forthcoming) Comparative Performance Measurement in Regulation: The Case of English and Welsh Sewerage Services, *Journal of the Operational Research Society*.

Thanassoulis E., (2000a) The Use of Data Envelopment Analysis in the Regulation of UK Water Utilities: Water Distribution, *European Journal of Operational Research.* Vol. 126, 2, pp. 436-453.

Thanassoulis E. (2000b) DEA and its Use in the Regulation of Water Companies, European *Journal of Operational Research*, Vol. 127, 1, pp. 1-13.

Thanassoulis E. (1999) Data Envelopment Analysis and its Use in Banking. *Interfaces* Vol. 29, No 3, pp. 1-13.

Thanassoulis E., (1997), Duality in data envelopment analysis under constant returns to scale, *IMA Journal of Mathematics Applied in Business and Industry*, Vol. 8, No. 3, pp. 253-266.

Thanassoulis E., (1995) Assessing Police Forces in England and Wales Using Data Envelopment Analysis, *European Journal of Operational Research,* Vol. 87, pp. 641-657.

Thanassoulis E. and R. Allen, (1998) Simulating weights restrictions in data envelopment analysis by means of unobserved DMUs. *Management Science*, Vol. 44, No. 4 pp. 586 - 594.

Thanassoulis E., A. Boussofiane and R. G. Dyson, (1995) Exploring Output Quality Targets in the Provision of Perinatal Care in England Using DEA.

European Journal of Operational Research, Vol. 80, No. 3 pp. 588-607.

Thanassoulis E. and R. G. Dyson, (1992) Estimating preferred input output levels using data envelopment analysis. *European Journal of Operational Research* Vol. 56 pp. 80-97.

Thompson, R. G., Singleton, F. D., Thrall, R.M. and Smith, B. A., (1986) Comparative Site Evaluations for Locating a High-Energy Physics Lab in Texas, *Interfaces,* Vol. 16, 6, pp. 35-49

Thompson, R. G.; Langemeier, L. N.; Lee, C.-H.; Lee, E.; and Thrall, R.M. (1990), The role of multiplier bounds in efficiency analysis with application to Kansas farming, *Journal of Econometrics,* Vol.46, pp. 93-108.

Thrall R. M. (2000) Measures in DEA with an Application to the Malmquist Index. *Journal of Productivity Analysis* Vol. 13, No. 2 pp. 125-137.

Winston W. L. (1994) *Operations Research: Applications and Algorithms* (3rd edition) Duxbury Press.

Wong Y-H.B. and J.E. Beasley (1990), Restricting Weight Flexibility in DEA. *Journal of the Operational Research Society,* Vol. 41, 9, 829-835.

Zhu J., (2000) Setting Scale Efficient Targets in DEA via Returns to Scale Estimation, *Journal of the Operational Research Society* Vol. 51, No. 3, pp. 376-378.

PUBLISHER DISCLAIMER